Critical Analysis of Prototype Autonomous Vehicle Crash Rates: Six Scientific Studies from 2015–2018

Critical Analysis of Prototype Autonomous Vehicle Crash Rates: Six Scientific Studies from 2015-2018

RICHARD A. YOUNG

Driving Safety Consulting, LLC

SAE INTERNATIONAL®

Warrendale, Pennsylvania, USA

400 Commonwealth Drive
Warrendale, PA 15096-0001 USA
E-mail: CustomerService@sae.org
Phone: 877-606-7323 (inside USA and Canada)
724-776-4970 (outside USA)
Fax: 724-776-0790

Library of Congress Catalog Number 2021938085
http://dx.doi.org/10.4271/9781468603422

ISBN-Print 978-1-4686-0341-5
ISBN-PDF 978-1-4686-0342-2
ISBN-epub 978-1-4686-0343-9

To purchase bulk quantities, please contact: SAE Customer Service

E-mail: CustomerService@sae.org
Phone: 877-606-7323 (inside USA and Canada)
724-776-4970 (outside USA)
Fax: 724-776-0790

Visit the SAE International Bookstore at books.sae.org

Chief Growth Officer
Frank Menchaca

Publisher
Sherry Dickinson Nigam

Development Editor
Publishers Solutions, LCC
Albany, NY

Director of Content Management
Kelli Zilko

Production and Manufacturing Associate
Erin Mendicino

To everyone who has suffered a loss from an automobile crash.

Contents

Table of Figures

Table of Tables

Foreword

Today we are witnessing rapid change in many of the large-scale systems of which we humans are a part. Nowhere is this more apparent than in our transportation system—where advanced technologies, including automation, are increasingly being integrated at many levels—with the hope (and aspiration) of improving safety, among other goals.

One of the critically important questions that has emerged about advanced technologies in transportation is how to test the actual effects of these advanced systems on safety—particularly how to evaluate the safety of highly automated driving systems. Many are asking whether a standard approach to safety verification and validation can be used or whether it requires augmentation or even a whole new approach. These questions have been tackled by any number of expert teams, but they are rife with challenging issues. Among these issues are the following: What tests and procedures should be included in a safety assurance process for automated driving systems? Can such a process include analytic evaluations and simulation testing in conjunction with a closed test track and public road testing? How should road testing be structured? Should it include *certain* scenarios? If so, *which scenarios are critical* for inclusion? How many scenarios are required? Should certain scenarios be prioritized? And, very importantly, for how many miles should an automated driving system be driven? Some states have made provisions for testing highly automated vehicles on their roadways under specific conditions, and at test facilities, thus enabling initial data to be collected on the experiences there.

Among various initiatives that have been taken to address some of these safety questions, this book focuses on the application of epidemiological methods to the evaluation of crash risk posed by automated driving systems—and, for several reasons, it is important to all of us working in the area of automation and safety.

***First*, this book challenges us to establish best practices for computing epidemiological evaluations of crash risk** and perhaps even to standardize the types of data that are used for them. More than that, it challenges us to think with care and precision about our application of all safety assurance methods. In Chapter 1, the book provides terminology and clear definitions from epidemiological approaches to crash risk assessment to help facilitate improved practices in this arena. This could be very important and helpful for advancing discussions around automation and methods for evaluating its safety.

***Second*, this book identifies key issues in cutting-edge papers from the published literature for the purpose of building a greater understanding of the elements of crash risk data and computations that are pivotal for accuracy and interpretation.** In Chapters 2, 3, 4, 5, 6, and 7, the book presents and reviews six pieces of recent research on crash risk for highly automated vehicles and has identified key issues (from an epidemiological perspective) that could be improved upon as we strive for more standardized, robust methods that can be applied to safety verification. In doing so, the book offers methodological specifics that are potentially a way to move the field forward toward standard methods of comparing data and performing computations so that studies can be compared and contrasted with one another meaningfully—and so that the field is better able to identify directions that future work should take. The book can thus serve as a resource and foundation for those concerned with working toward a vision of improved safety in the future. It is by thoroughly reviewing the first papers published on crash risk for automated driving systems—and by identifying and correcting potential biases and moving the computations toward best practices in epidemiology—that the book strives to establish an "improved" foundation of reviewed work that has already incorporated best-practice methods within it. It is one that the field could now use to build upon as new studies are reported and contribute to a growing understanding of how automation affects crash risk. In addition, once practitioners and researchers are aware of potential pitfalls and areas of bias in performing epidemiological computations, and understand them, they are in a position to decide if they agree or disagree. But without this awareness, we in the field are less prepared than we may realize to think about risk from an epidemiological perspective. This book strives to equip us with sensitivity, knowledge, and foundation that can be used to build upon as more data accumulates (and as the field harnesses some of the lessons it provides).

***Third*, this book provides a useful compilation in one place** about the epidemiology of crash risk computations tailored to driving automation systems, including terminology, a review of computations, and a description of different types of biases for which new data and new computations might be checked when they are being undertaken. The main chapters of the book provide this information—together with a summary in Chapters 8, 9, and 10, and then recommendations at the end in Chapter 11.

In closing this Foreword, let me say that in the effort to develop a solid and robust set of methods for evaluating safety risk of new technologies in the vehicle—*precision* and *care* in the application of concepts is vital. And this is indeed the heart of where Dr. Young has a role to play. I first met Dr. Young in the late 1990s, in the context of a similar challenge that arose from a different technology. At that time, we were trying to develop a framework for evaluating the effects of new technology on driver workload, attention, and distraction. Then, as now, one question that confronted us was how technologies affected crash risk. From the work that we were doing at that time until now, it has been clear that Dr. Young cares very deeply about safety, is willing to tackle even the most difficult of technical problems, willing to look at objective evidence with a sharp and clear eye, and willing to speak to the facts, whether it is to articulate evidence that is consistent with safe use or whether to ask tough questions—even when this means asking those questions of highly respected colleagues or, sometimes, challenging the field as a whole to think more carefully. Whether you agree or disagree with an approach or viewpoint taken by Dr. Young, you come away from an interaction with him—or a paper of his—knowing without a doubt that he has pushed you to think a little bit harder or that he has led you to think about things differently and that he has challenged you to think perhaps more clearly and deeply than you had before. And that is something vital to the field of safety and automation right now—whether building theory and methods or applying them in practice. Moreover, for more than 20 years now, Dr. Young has acquired an in-depth mastery of epidemiology applied to safety science—seeking out consultation from leading experts in epidemiology—and learning through interactions with them over those two decades. As a result of both his expertise and commitment to safety, he is particularly well suited to have done the work in this book.

Of course, to reduce the crashes that continue to claim lives on our world's trafficways, much can be done in many areas. And accomplishing this will require harnessing multiple scientifically supported methods from related disciplines. But in this effort, this book can play an important role—as we all join in applying the best information available, augmenting and modifying solutions as new learning is gained, using it toward the effort to save more lives, further reduce injuries, and work toward an improved transportation system in the future.

Linda S. Angell

Linda S. Angell, Ph.D.
President and Principal Scientist
Touchstone Evaluations, Inc.
Detroit-Metro-Area, Michigan, USA

Preface

Will automated vehicles (AVs) be a net safety benefit or harm compared to human-driven conventional vehicles (CVs)? The answer is critical for society. This book attempts a preliminary answer by reviewing and then critically analyzing the first six scientific studies of AV crash rates. These six studies, published from 2015 to 2018, investigated whether the crash rates of prototype AVs with a backup safety driver were lower or higher than the crash rates of CVs. Four studies concluded that AVs *increased* the crash rate compared to CVs, while the two other studies concluded that AVs *decreased* it, sometimes after examining almost identical AV data. This book attempts to resolve this discrepancy by critically analyzing each study to reduce systematic errors, biases, and best practice violations.

A substantial bias often arose from CV crashes unrecorded by police. Several studies selected *all* AV crashes (police recorded and unrecorded) but *only* police-recorded crashes. This unrecorded crash selection bias falsely increased the AV crash rate compared to the CV crash rate. Site selection bias arose when an AV crash rate on city streets was compared to the national CV crash rate on all roads. Best practice violations occurred from combining heterogeneous crash severities or combining heterogeneous data across AV developers. In three of the six studies, this book corrected the errors, removed the biases, and avoided heterogeneity best practice violations by stratifying crash severity. However, the final corrected, adjusted, and stratified AV IRR estimates in these three studies were inconsistent. One AV IRR estimate was lower, another similar, and the third higher than the CV IRR of one, indicating unknown remaining biases.

The remaining biases were removed by analyzing AV and CV crash rate *trends*. The upward trend in the Google AV cumulative crash rate from 2015 to 2016 was explained by (1) an upward trend in the CV cumulative crash rate in Mountain View, CA, and (2) a downward trend in the Google ADS and test driver cumulative disengagement rate (the rate of transitions from autonomous mode to manual mode).

The study with the fewest biases and errors compared the crash rates of Google prototype AVs to the crash rates of naturalistic driving study CVs. After using standardization to adjust for site selection bias, Google AVs reduced Severe crashes by 45%, about the same as CVs fully equipped with crash avoidance technologies. However, Google AVs had little effect on Major or Minor Property Damage Only crashes. Additional AV crash and mileage data are required to improve the validity and precision of these estimates.

Author's Note Regarding Abbreviations, Keywords, and Phrases

This book combines concepts and terminology from three fields: automotive safety, epidemiology, and automated vehicles. Many new phrases were coined by combining words and ideas from these fields. For example, the phrase "unrecorded crash selection bias" combines the term "unrecorded crash" from automotive safety and the term "selection bias" from epidemiology. When reading this book, it is helpful to recognize that this string of words is one phrase with a specific definition.

Many such new phrases were coined and used throughout this book. Many of these phrases have specific technical meanings that will not be familiar because they are novel. Hence, the phrases are listed below to call attention to them in a "meta" way to help you be more aware of them as you read the book. The goal is to increase awareness that these phrases are being used in a specific manner.

Each entry below is defined in Appendix B on page 149, which you may want to read first. In the e-book, each entry below also links to its definition in Appendix B.

ADS
APM
automated driving system
autonomous
AV
autonomous mode
autonomous vehicle
bench report
best practice violation
bias
CA
CHP
CI
Conditional Driving Automation
confidence interval
confidence limits
conventional mode
conventional vehicle
crash rate
crash selection bias
crash Severity level
crash severity level
cumulative crash rate
cumulative disengagement rate
CV
DDT
Delta-V
disengagement
DMV
driving automation system
dynamic driving task
environment selection bias
FHWA
FMCSA
GM
GPS
HAPM
heterogeneity
HPMS
incidence rate
incidence rate ratio
IRR
Level 1 crash
Level 2 crash
Level 3 crash
limitation
LL
Major PDO
manual mode
Median APM
Median DPM
mid-*P*
Minor PDO
MVMT
naturalistic driving study
NDS
NHTSA
non-PR crash
ODD
operational design domain
PAR

PDO
Police Accident Report
police record
police-recorded crash
police report
police-reportable crash
police-reported crash
police-unrecorded crash
PR
PR crash selection bias
road selection bias
SAE
SAE Level 3
selection bias
self-driving
Severe crash
severe crash
Severity Level
severity level
SHRP2
sIRR
site selection bias
standardization
standardized incidence rate ratio
SWITRS
systematic error
take over
take-over request
test driver
testing
TIMS
total crash rate
total disengagement rate
transition
transition crashes
UL
UMTRI
unrecorded crash
unrecorded crash selection bias
unreported crash
unreported crash selection bias
VMT
VTTI

Acknowledgments

For helpful comments on an early draft, I thank (in alphabetical order) Subho Banerjee, Sai Chang, Vinayak Dixit, Brandon Schoettle, Chris Schreiner, Larry Smythe, Jing Zhang, and other colleagues who prefer to remain anonymous. I especially thank my Senior Research Associate, Amanda Zeidan. She independently matched AV crashes to police crash databases, formatted the initial manuscript into SAE International's book format, updated the reference citations, and proofread the final manuscript. I thank my wife Alice for family support. Finally, I thank SAE for editing and formatting the final manuscript. At the beginning of each chapter, I summarize the original data in each of the six studies. However, I alone am responsible for the additional analyses that correct errors, adjust for biases, and estimate the final AV crash incidence rate ratios. I received no outside funding for this research and have no declarations of interest.

Introduction

Are autonomous vehicles (AVs)[1] a net safety benefit or harm compared to human-driven conventional vehicles (CVs)? The first answers to this question were based on "hopes" or modeling studies (see Appendix A) without empirical data from real-world AV crashes. However, six scientific studies of AV crashes from 2015 to 2018 were the first to be based on AV empirical data. This book critically reviews the data and methods in each of these six studies. It first presents the original data in each study in an epidemiologic format. It then identifies and corrects the systematic errors, biases, and best practice violations to the extent possible. Finally, it reanalyzes the updated data using epidemiologic methods.

Specifically, this book reviews the first six journal articles or academic reports that compared the crash rate of prototype AV fleets (with a backup professional test driver[2]) to CV crash rates. These studies are critically analyzed in their order of publication. Each study is first summarized using its original data and analysis. Next, each study's original data are reanalyzed using epidemiologic methods rather than conventional statistics. These epidemiologic methods helped identify critical issues in each study, such as systematic errors, biases, and best practice violations. The data and methods in each study were then corrected where necessary, the biases adjusted, and the best practice violations avoided wherever possible. Next, the updated data were reanalyzed, and the revised results were

---

[1] The abbreviation "AV" is used in this book as an informal abbreviation for the formal definition in the Society of Automotive Engineers J3016 standard as an SAE "level 3 ADS-operated vehicle" (https://www.sae.org/standards/content/j3016_202104). The abbreviation AV used in this book also refers specifically to prototype AVs with a backup test driver, not production AVs intended for use by the general public. AVs are also often informally termed self-driving vehicles or "automated vehicles." See Appendix B, "Abbreviations and Definitions," for further explanation.

[2] Terms in *italic* in this book are defined in Appendix B.

presented for each study. Extensive Appendices give details about the data and methods and raise additional issues pertinent to AV crash safety.

I reached a preliminary conclusion about AV safety based on the only study with no issues remaining after further analysis (Chapter 3). The Google prototype AV fleet in the study period tended to reduce the Severe crash rate compared to CVs but had a negligible effect on Major and Minor Property Damage Only (PDO) crashes.

However, this book's main objective is not to show that these first six scientific studies had errors and biases (all studies do). The main objective is also not to arrive at a conclusion about AV safety from the early data from prototype AVs. Instead, the main objective is to help future studies analyze real-world AV safety data and avoid the systematic errors, biases, and best practice violations identified in this book. Only then can future studies arrive at valid conclusions about AV safety based on real-world data.

1

Methods

Contents

Note: Refer to Appendix B on page 149 for an explanation of the abbreviations and the definitions of unique terms and phrases.

The six journal articles and scientific reports reviewed in this book (and the critical analysis of the data and methods in those articles and reports) use terms from driving safety and epidemiology that may not be familiar to some readers. Therefore, Appendix B lists definitions of critical driving safety and epidemiologic acronyms and terms used in this book.

Chapter 1 describes the following methods. First, the epidemiologic method used to estimate the incidence rate ratio (IRR) is introduced. Second, the method for estimating the confidence limits of the IRR is presented. Third, the method used to test the heterogeneity of the IRR for different groups is given. Fourth, the method for matching AV crashes to police records is described. Finally, the reasons are given for avoiding statistical significance tests in the data analyses in this book, as per the current best practice in epidemiology.

1.1. IRR Estimation Method

Designs for epidemiologic metrics and their methods for application to driving safety issues are described by Young (2018) [1.1]. All the studies in this review compared an estimated AV crash rate to an estimated CV crash rate. The epidemiologic design is a rate ratio cohort design with driving-time as the denominator (see Section 1.1 in Young (2018) [1.1]). After correcting issues in the original data of the six reviewed studies, this review then compares the corrected AV and CV crash rates using the epidemiologic IRR metric and method.

The IRR metric is useful for understanding crash risks because it offers a well-established scientific metric and method to evaluate exposure risks in driving safety (Young, 2018,

TABLE 1.1 Epidemiologic method to estimate the IRR for AV exposure.

	AV exposed	Not AV exposed
Crashes	B	E
MVMT	C	F
Crash rate	B/C	E/F
IRR	(B/C) (E/F) = BF/EC	

Section 1.1) [1.1]. Dividing the number of crashes by miles traveled, the resulting crash rate estimate is in units of crashes per miles traveled. Or, more commonly, the units are crashes per Millions of Vehicle Miles Traveled (MVMT). The AV crash IRR is estimated simply by dividing the AV crash rate estimate by the CV crash rate estimate. Table 1.1 illustrates the epidemiologic IRR estimate method from Young (2018, Section 1.1) [1.1]. This method is applied throughout this book to estimate the relative crash rates for AV exposure versus non-exposure.

Table 1.1 compares the crash rates for "AV exposed" versus "Not AV exposed." ("Not AV exposed" in these studies is the same as "CV.") Table 1.1 (left column in 2×2 matrix) is the data when vehicles were exposed to the autonomous mode in the subset of the vehicles in the defined population under investigation. First, the number of crashes while exposed to autonomous mode is given by cell (B). Next, the number of autonomous mode crashes is divided by the total miles traveled in autonomous mode (C). (In this book, mileage is always given in MVMT units unless otherwise indicated.) The ratio (B/C) is the crash rate when a vehicle is exposed to the autonomous mode (i.e., the AV crash rate).

Table 1.1 (right column in 2×2 matrix) gives the data for vehicles not exposed to autonomous mode (i.e., human-driven CVs). First, the number of CV crashes that occurred is given by cell (E). Next, the number of CV crashes is divided by the total CV MVMT (F). The ratio (E/F) is then the crash rate estimate when not exposed to autonomous mode (i.e., the CV crash rate estimate).

The AV IRR estimate is then the ratio of the "AV exposed" (B/C) to the "Not AV exposed" crash rate estimates (E/F). In other words, the AV IRR is an estimate of the population IRR for autonomous exposure versus non-exposure. Simply put, the AV IRR estimate in the population under study is the AV crash rate estimate divided by the CV crash rate estimate.

The IRR is a dimensionless ratio. If the AV crash rate estimate is higher than the CV crash rate estimate, then the AV IRR estimate is higher than one, and the AV effect is said to be causative of crashes. If the AV crash rate is lower than the CV crash rate, then the AV IRR estimate is less than one, and the AV effect is said to be *protective* of crashes. The percentage change in the AV versus CV crash rate is given by the equation 100 *x* (IRR – 1).

Note also that the IRR, although it is a measure of relative rates, has about the same numerical value as the relative risk (see Young, 2018) [1.1]. The effect size estimates of the risk ratio and rate ratio should be about the same. As stated by Young (2018, p. 6) [1.1], "Different study designs of a risk factor, in the absence of bias, should give rise to about the same effect size."

1.2. IRR Confidence Limit Estimation Method

There are multiple epidemiologic methods to estimate IRR confidence limits. The choices were narrowed down for this study to the following. First, an exact method based on dual Poisson distributions rather than an approximate method based on dual Gaussian distributions was necessary to provide an accurate probability value when the AV-exposed cases are small (Rothman et al., 2008, p. 253) [1.2]. (The definition of "small" varies in the epidemiologic literature from less than about 6 to less than about 80.) The California (CA) AV test data examined in the six studies under review had a small number of crashes when in autonomous mode or during a transition, requiring the use of "dual Poisson" rather than "dual Gaussian" distributions. The small-sample methods described in (Rothman et al., 2008, p. 254) [1.2] provide an exact analysis of a 2×2 table (such as Table 1.1) when the crash counts are small. These methods are well suited to estimate small-sample probability values and exact IRR confidence limits. They are adapted here for driving safety studies with crash and mileage data in the Table 1.1 design. Second, the mid-P exact confidence limit method was used rather than the "Fisher exact confidence limit" method because the mid-P method gives rise to a smaller CI.

Although AV crash counts are small, the CV crash count at the United States (U.S.) national level is large (in the thousands for fatal crashes and millions for total crashes). Few epidemiologic programs can estimate exact 95% confidence limits with only a few exposed cases and many unexposed cases. Of the two, the StatXact [1.3] program "CI on common relative risk" gives the mid-P exact confidence limits but not the mid-P probability; it instead provides the Fisher probability. The other program (Stata [1.4] EpiTab, "iri") gives the mid-P probability, but not the mid-P confidence limits; it instead provides the Fisher confidence limits. Therefore, the StatXact [1.3] program was used to estimate the mid-P confidence limits, and the Stata [1.4] program was used to estimate the mid-P probability.

The StatXact mid-P exact confidence limits and the Stata mid-P probability estimates were confirmed with two other programs comparing two rates with a dual Poisson model. These were OpenEpi [1.5] ("compare two rates" program) and the R program for statistical computing [1.6] ("rateratio" [1.7] program in the library "epitools" [1.8]). R programs can be run on the web directly [1.9]. However, these two programs only worked without error when there were less than several hundred CV crashes.

The StatsDirect [1.10] ("compare two crude rates") also used a dual Poisson model and was successful even with CV crashes up to about 20,000. If CV crash counts were more than 20,000, the crashes and miles could both be divided by 1,000, and the StatsDirect [1.10] IRR, confidence limits, and mid-P probability estimates were used to check for consistency with those from StatXact [1.3] and Stata [1.4].

Note also that the StatXact [1.3] and Stata [1.4] programs can only process event counts that are integers of zero or above. After standardization, as described in Section 3.4, event counts are typically non-integers (i.e., they have a decimal component). In such instances, the StatsDirect [1.10] program was used, which has the added benefit of providing mid-P confidence limits and mid-P probability values in a single program. Therefore, throughout this book, the final IRR, mid-P confidence limits, and P-values were estimated using StatsDirect [1.10], unless indicated otherwise.

1.3. Heterogeneity Best Practice Violation

Heterogeneity is defined in the field of epidemiology as "Differences in stratum-specific effect measures. When such measures are not equal, it is said that the effect measure is heterogeneous or modified or varies across strata." (Porta, 2014, p. 134) [1.11]. Pooling heterogeneous strata hides important properties of the individual strata. Therefore, the best practice in epidemiology is to detect and estimate differences between strata (also known as "effect measure modification") and report those as findings, not hide them by collapsing (i.e., summing or combining) across strata (Rothman et al., 2008, p. 259) [1.2]. For example, crashes at different severity levels are often heterogeneous (Knipling, 2017 [1.12]; Knipling and Wåhlberg, 2019 [1.13]). Therefore, pooling crashes across severity levels (for any combination of fatal, severe injury, minor injury, and property damage crashes) without first testing for homogeneity is a best practice violation.

In this book, the heterogeneity (between crash Severity Levels and the crash data from different AV developers) was assessed using the meta-analysis tab in the Episheet (2015) program [1.14]. An IRR of zero was assigned a value of 0.001 for purposes of heterogeneity testing because zero has an indeterminate logarithm, and the Episheet meta-analysis program returns "invalid entry." Homogeneity was verified when the number of crashes was an integer of zero or greater using the "Rate Data" program in Episheet (2015) [1.14]. If the data were homogeneous (across crash severity strata or AV developer strata), then the total IRR and its exact confidence limits were estimated based on the AV and CV crash totals across the strata. The AV mileage is the same for each AV crash Severity Level, and the total AV crashes could then be validly summed across all Severity Levels. Likewise, the CV mileage is the same for each CV crash severity level, and the total CV crashes could be summed across all severity levels.

If the data were homogeneous, then the estimates of the total crash IRR, its confidence limits, and its P-value were based on the "crude" crash and mileage estimate rather than the Mantel-Haenszel "pooled" IRR estimate. This limitation is because the Mantel-Haenszel method sums the mileage data across the severity strata, falsely tripling the mileage in the pooled data, causing a downward bias in the confidence interval (CI).

In some conditions where AV crash counts were small, the heterogeneity between strata (whether crash severity strata or company strata) was checked using exact methods with the StatsDirect program [1.10]. The meta-analysis program for IRRs in StatsDirect [1.10] used a pooled fixed effects model. The goal was to use the StatsDirect program to confirm the more straightforward heterogeneity test in the Episheet [1.14] meta-analysis program. If the Episheet test indicated heterogeneity, the StatsDirect generally confirmed that result (that is, the StatsDirect meta-analysis Cochran Q test rejected the null hypothesis that the strata had similar IRRs, and therefore could not be validly combined). In other words, the StatsDirect program [1.10] was used to confirm the heterogeneity result given by the Episheet program [1.14]. However, with colossal CV crash numbers, such as at the national level, the StatsDirect program [1.10] sometimes returned an error message of "Array dimensions exceeded supported range." In such cases, the crash and mileage numbers were divided by some power of 10, which does not affect the crash rate, and the StatsDirect program [1.10] was re-run to confirm the heterogeneity.

1.4. Matching AV Crashes to Police Records

A new method was developed to avoid arbitrary assumptions about the percentages of unreported or unrecorded CV injury and property-damage-only (PDO) crashes (Appendix E). The new method directly compares police-recorded AV crashes to police-recorded CV crashes to look for a match.

To attempt the match, my research associate and I independently searched the California Highway Patrol Statewide Integrated Traffic Records System (CHP SWITRS) [1.15] crash database records to find a match to an AV crash in the CA police database. The matches were based on details in the crash data and descriptions in Blanco et al. (2016) [1.16], Google Reports [1.17], and California Department of Motor Vehicles (CA DMV) reports [1.18]. Matches were made by date, approximate time (within one hour), approximate location (within 500 feet according to the global positioning system [GPS] coordinates), type of crash (rear-end, sideswipe, etc.), and the make and model of the base vehicle model for the AV as specified by the AV company. The last criterion (matching the make and model) was accomplished using the SWITRS "party file" downloaded from the Transportation Injury Mapping System interface [1.19] to SWITRS [1.15]. This final match of the make and model is critical because sometimes a match could be found with several of the first criteria but fail the last "make and model" match criterion.

Searches were performed in the CA police record SWITRS database [1.15] for all CA AV crashes investigated in all six studies at all crash Severity Levels 1–3 (Severe, Major PDO, and Minor PDO). See Appendix G for crash Severity Level definitions.

1.5. Reason for Not Using Statistical Significance Tests

As per the current best practice in epidemiology and statistics, the present reanalysis of these six studies gives no statistical significance tests for the reanalyzed data. What is instead presented is the point estimate of each AV crash IRR, its confidence limits, its P-value, and the percentage increase or decrease of the AV crash rate compared to the CV crash rate for the IRR point estimate and its confidence limits.

The reason for not using statistical significance tests in this book is that, over the last forty years, there has been a growing consensus that such tests should not be used in scientific publications. It is impossible to cite all the relevant articles here; there are too many. Essential highlights are noted below.

The current best practice in epidemiology categorically rejects the use of statistical "significance" criteria to make pass-fail decisions based on a probability value (e.g., "statistically significant at $P < 0.05$") (Rothman, 1978 [1.20]; Gardner and Altman, 1986 [1.21]; Altman and Bland, 1995 [1.22]; Rigby, 1999 [1.23]; Greenland et al., 2016 [1.24]; Amrhein et al., 2019 [1.25]).

Rothman (1986) [1.26] writes, "Testing for significance continues today not on its merits as a methodological tool but on the momentum of tradition. Rather than serving as a thinker's tool, it has become for some a clumsy substitute for thought, subverting what should be a contemplative exercise into an algorithm prone to error." Rothman (1998) [1.27] further states, "Despite a widespread belief that many journals require significance tests for publication, the Uniform Requirements for Manuscripts Submitted to Biomedical Journals discourages them, and every worthwhile journal will accept articles that omit them entirely. In Epidemiology, we do not publish them at all."

The rejection of statistical significance testing has also been adopted as the best practice not just in Epidemiology but also in Statistics (e.g., Berger and Delampady, 1987 [1.28]; Altman et al., 2013 [1.29]; Greenland et al., 2016 [1.24]; Wasserstein and Lazar, 2016 [1.30]). For example, Altman et al. (2013) [1.29] note that "It is clear that there is a considerable consensus among statisticians that confidence intervals represent a far better approach to the presentation and interpretation of results than significance tests and P values." The journal *Basic and Applied Social Psychology* bans P-values entirely for null hypothesis significance testing

(Trafimow and Marks, 2015) [1.31]. The official position of the American Statistical Association (Wasserstein and Lazar, 2016) [1.30] is that "The widespread use of 'statistical significance' (generally interpreted as '$p \leq 0.05$') as a license for making a claim of a scientific finding (or implied truth) leads to considerable distortion of the scientific process." Wasserstein et al. (2019) [1.32] state, "We conclude, based on our review of the articles in this special issue and the broader literature, that it is time to stop using the term 'statistically significant' entirely. Nor should variants such as 'significantly different,' '$p < 0.05$,' and 'nonsignificant' survive, whether expressed in words, by asterisks in a table, or in some other way."

References

1.1. Young, R.A., "Epidemiological Study Designs Applied to Driving Safety: Definitions and Examples," SAE Technical Paper 2018-01-0496, 2018, https://doi.org/10.4271/2018-01-0496, https://www.researchgate.net/publication/324190638, accessed May 19, 2021.

1.2. Rothman, K., Greenland, S., and Lash, T., *Modern Epidemiology*, 3rd ed. (Philadelphia, PA: Lippincott Williams & Wilkins, 2008).

1.3. StatXact, Epidemiology Software, Version 12, January 5, 2016, https://www.cytel.com/software/statxact, accessed September 22, 2021.

1.4. Stata, "Statistics/Data Analysis," Computer Software, Rev. 13.1, StataCorp, College Station, TX, December 16, 2016, http://www.stata.com, accessed May 19, 2021.

1.5. OpenEpi, Epidemiology Software, Version 3.01, April 6, 2013, https://www.openepi.com/PersonTime2/PersonTime2.htm, accessed May 19, 2021.

1.6. "R Program for Statistical Computing," Statistics Software, https://www.r-project.org/, accessed May 19, 2021.

1.7. "Rateratio: Calculate Incidence Rate Ratio and Its Confidence Intervals," Epidemiology R Software, https://www.rdocumentation.org/packages/fmsb/versions/0.7.1/topics/rateratio, accessed May 19, 2021.

1.8. "epitools: Epidemiology Tools," Epidemiology Software, https://rdrr.io/cran/epitools/, accessed May 19, 2021.

1.9. "R Statement Program Interface," Computer Software for Epitools, https://cran.r-project.org/web/packages/epitools/index.html, accessed May 19, 2021.

1.10. StatsDirect, Statistical Software, Version 3.2.8, September 16, 2019, https://www.statsdirect.com/, accessed September 22, 2021.

1.11. Porta, M., *A Dictionary of Epidemiology*, 6th ed. (New York: Oxford University Press, 2014), 343 pp., pp., http://irea.ir/files/site1/pages/dictionary.pdf, accessed May 19, 2021.

1.12. Knipling, R.R., "Crash Heterogeneity: Implications for Naturalistic Driving Studies and for Understanding Crash Risks," *Transportation Research Record* 2663(1) (2017): 117–125, doi:10.3141/2663-15, https://journals.sagepub.com/doi/abs/10.3141/2663-15, accessed May 19, 2021.

1.13. Knipling, R.R. and Wåhlberg, A.E., "The Heterogeneity Principle," in *Proceedings of the Tenth International Driving Symposium on Human Factors in Driver Assessment, Training and Vehicle Design*, University of Iowa Public Policy Center, Santa Fe, NM, 2019, 224–230, doi:10.3141/2663-15, https://drivingassessment.uiowa.edu/sites/drivingassessment.uiowa.edu/files/da2019_35_knipling_final_0.pdf, accessed May 19, 2021.

1.14. Rothman, K.J., "Episheet: Spreadsheets for the Analysis of Epidemiologic Data," 2015, https://www.colleaga.org/tools/episheet, accessed May 19, 2021.

1.15. SWITRS, "California Highway Patrol Statewide Integrated Traffic Records System," https://iswitrs.chp.ca.gov/Reports/jsp/index.jsp, accessed May 19, 2021.

1.16. Blanco, M., Atwood, J., Russell, S., Trimble, T. et al., "Automated Vehicle Crash Rate Comparison Using Naturalistic Data," Virginia Tech Transportation Institute, January 2016, https://vtechworks.lib.vt.edu/handle/10919/64420, accessed May 19, 2021.

1.17. Google Reports, "Google Self-Driving Car Project Monthly Reports," Mountain View, CA, May 2015–November 2016. Note: Monthly reports are longer publicly available from Google. However, 19 monthly Google Reports from May 2015 to November 2016 are available at https://github.com/wgetsnaps/ca-dmv-autonomous-vehicle-accidents/tree/master/google-pdfs, accessed May 19, 2021.

1.18. CA DMV, "Report of Traffic Collision Involving an Autonomous Vehicle (OL 316)," California Department of Motor Vehicles, 2014–2019, https://www.dmv.ca.gov/portal/uploads/2021/05/OL316.pdf, accessed August 14, 2021.

1.19. TIMS, "Transportation Injury Mapping System," https://tims.berkeley.edu/tools/query/summary.php, accessed May 19, 2021.

1.20. Rothman, K.J., "A Show of Confidence," *The New England Journal of Medicine* 299 (1978): 1362–1363, doi:10.1056/NEJM197812142992410.

1.21. Gardner, M.J. and Altman, D.G., "Confidence Intervals Rather Than P Values: Estimation Rather Than Hypothesis Testing," *Br Med J (Clin Res Ed)* 292, no. 6522 (1986): 746-750, doi:10.1136/bmj.292.6522.746, accessed May 19, 2021.

1.22. Altman, D.G. and Bland, J.M., "Statistics Notes: Absence of Evidence Is Not Evidence of Absence," *BMJ* 311, no. 7003 (1995): 485, doi:10.1136/bmj.311.7003.485, accessed May 19, 2021.

1.23. Rigby, A.S., "Getting Past the Statistical Referee: Moving away from P-Values and towards Interval Estimation," *Health Education Research* 14, no. 6 (1999): 713–715, https://academic.oup.com/her/article/14/6/713/745885, accessed May 19, 2021.

1.24. Greenland, S., Senn, S.J., Rothman, K.J., Carlin, J.B. et al., "Statistical Tests, P Values, Confidence Intervals, and Power: A Guide to Misinterpretations," *European Journal of Epidemiology* 31, no. 4 (2016): 337–350, https://link.springer.com/article/10.1007/s10654-016-0149-3, https://pubmed.ncbi.nlm.nih.gov/27209009/, accessed May 19, 2021.

1.25. Amrhein, V., Greenland, S., and McShane, B., "Scientists Rise Up against Statistical Significance," *Nature* 567 (2019): 305–307, doi:10.1038/d41586-019-00857-9, https://www.nature.com/articles/d41586-019-00857-9, accessed May 19, 2021.

1.26. Rothman, K.J., "Significance Questing," *Annals of Internal Medicine* 105, no. 3 (1986): 445–447, doi:10.7326/0003-4819-105-3-445, https://annals.org/aim/article-abstract/700736/, accessed May 19, 2021.

1.27. Rothman, K.J., "Writing for Epidemiology," *Epidemiology* 9, no. 3 (1998): 333-337, https://journals.lww.com/epidem/Citation/1998/05000/Writing_for_Epidemiology.19.aspx, accessed May 19, 2021.

1.28. Berger, J.O. and Delampady, M., "Testing Precise Hypotheses," *Statistical Science* 2 (1987): 317–335, doi:10.1214/ss/1177013238, https://projecteuclid.org/download/pdf_1/euclid.ss/1177013238, accessed May 19, 2021.

1.29. Altman, D., Machin, D., Bryant, T., and Gardner, M., *Statistics with Confidence: Confidence Intervals and Statistical Guidelines*, 2nd ed. (New York: John Wiley & Sons, 2000), 321 pp.

1.30. Wasserstein, R.L. and Lazar, N.A., "The ASA's Statement on p-Values: Context, Process, and Purpose," *The American Statistician* 70, no. 2 (2016): 129–133, doi:10.1080/00031305.2016.1154108, https://amstat.tandfonline.com/doi/full/10.1080/00031305.2016.1154108, accessed May 19, 2021.

1.31. Trafimow, D. and Marks, M., "Editorial in Basic and Applied Social Psychology," *Basic and Applied Social Psychology* 37 (2015): 1–2, doi:10.1080/01973533.2015.1012991, https://www.tandfonline.com/doi/full/10.1080/01973533.2015.1012991, accessed May 19, 2021.

1.32. Wasserstein, R.L., Schirm, A.L., and Lazar, N.A., "Moving to a World Beyond 'p< 0.05'," *The American Statistician* 73, no. sup1 (2019): 1–19, doi:10.1080/00031305.2019.1583913, https://www.tandfonline.com/doi/full/10.1080/00031305.2019.1583913, accessed May 19, 2021.

2

The First Major AV Safety Study: Schoettle and Sivak (2015)

Contents

Note: Refer to Appendix B on page 149 for an explanation of the abbreviations and the definitions of unique terms and phrases.

Schoettle and Sivak (2015) [2.1], automotive researchers at the University of Michigan Transportation Research Institute (UMTRI), published the first scientific study of publicly available AV safety data.

2.1. Summary of Schoettle and Sivak (2015)

Schoettle and Sivak (2015) [2.1] estimated the AV crash rates for three of the ten AV companies that had been approved for public road testing in CA by the time of their 2015 study. The three companies were Google, Delphi, and Volkswagen (VW)/Audi.

2.1.1. Original Schoettle and Sivak (2015) Data: AV versus CV Crash Rates

The UMTRI investigators counted crashes during the transition from autonomous mode to manual mode as in the same group as "autonomous" crashes. They did so because the critical events "developed while the vehicles were in "autonomous mode," and there is no indication that the outcomes would have been different if the vehicle had remained in "autonomous mode." (See Appendix D for crash descriptions and comments about transition crashes.)

Schoettle and Sivak (2015) [2.1] summed the AV crash data and the AV mileage data across Google, Delphi, and VW/Audi. There were 11 AV crashes from December 2009 to September 30, 2015 (as per Schoettle and Sivak, 2015, Table 2) [2.1], including two crashes during the transition from autonomous to manual mode, which they counted as AV crashes. All 11 AV crashes were Google AV crashes in Mountain View, CA, and vicinity, with a total of 1.214626 million vehicle miles traveled (MVMT) in autonomous mode (Schoettle and Sivak, 2015, footnote 6) [2.1]). The mileage included: Google AV mileage in the Mountain View, CA area and Austin, TX; Delphi highway mileage across many states; and VW/Audi mileage in California and Nevada. Although the crash data were from 2012 to 2015, the mileage data for Google extended back to 2009 (Google Reports, May 2015) [2.2]. (Google did not report any AV crashes from 2009 to 2011.) Table 2.1 (left data column) shows Schoettle and Sivak's (2015) [2.1] original data comparing the AV crash rate to the 2013 U.S. CV crash rate.

TABLE 2.1 Schoettle and Sivak (2015): Original data.

	AV[a]	2013 U.S. CV[e]
Crashes	11[b]	5,687,057[f]
MVMT[i]	1.214626[c]	**2,988,323[g]**
Crash rate	9.1[d]	1.9[h]

Crashes, mileage, and crash rates for AVs and 2013 U.S. CVs.

Crashes are summed across severities (fatal + injury + PDO).

A **bold** number in the 2×2 data matrix above indicates no identified error or bias.

The underlined words in the notes below are revised in later tables.

Notes:

[a] AV data from December 2009 to September 30, 2015, for Google (Google Reports [2.2]), Delphi, and VW/Audi.

[b] All AV crashes, nine in autonomous mode and two in transition (Schoettle and Sivak, 2015, p. 11, footnote 6) [2.1].

[c] All AV miles summed across companies Google, Delphi, and VW/Audi (Schoettle and Sivak, 2015, p. 11, footnote 5) [2.1]. Includes non-CA mileage.

[d] AV crash rate.

[e] 2013 U.S. national CV data (NHTSA, 2015, p. 2) [2.3].

[f] Police-recorded crashes only.

[g] Estimated MVMT on all U.S. roads.

[h] 2013 U.S. CV police-recorded crash rate estimate.

[i] Million Vehicle Miles Traveled.

Table 2.1 (left data column) shows the total AV crash and mileage data across the three prototype AV developer companies. These data estimate an AV crash rate of 9.1 crashes per MVMT (note d).

Table 2.1 (right data column) shows the 2013 U.S. national CV crash and mileage data (NHTSA, 2015) [2.3], which estimate a CV crash rate of 1.9 crashes per MVMT (note h).

The AV data in Table 2.1 are all crashes and miles traveled in autonomous mode as recorded by the AV instrumentation. However, the 2013 U.S. CV data in Table 2.1 are only estimates, based only on samples of police records for crashes and road sensor measurements by state governments for mileage. Schoettle and Sivak (2015, p. 11) [2.1] initially state from their data in Table 2.1 that the overall crash rate for AVs is nearly five times the police-recorded crash rate for U.S. CVs. Consistent with their five-times ratio estimate, the IRR estimate from Table 2.1 is calculated as 9.1 ÷ 1.9 = 4.8.

2.1.2. Unrecorded Crash Selection Bias: CV Crashes (Table 2.1, Note f)

Schoettle and Sivak (2015) [2.1] correctly recognized a bias in the method they used to select the CV crashes in Table 2.1. For CVs, only police-recorded crashes were selected (Table 2.1, note f). For AVs, all crashes (police-recorded and police-unrecorded) were selected (Table 2.1, note b). The five times larger AV crash rate estimate of 9.1 in Table 2.1 (note d) compared to the CV crash rate estimate of 1.9 (note h) was therefore biased high because of the differential selection criteria for the AV and CV crashes (all crashes versus police-recorded crashes). This type of bias is known as selection bias, defined in Epidemiology as "Bias in the estimated association or effect of an exposure on an outcome that arises from the procedures used to select individuals into the study or the analysis." (Porta, 2014, p. 258) [2.4]. Here, the bias specifically concerns the differential criteria used for selecting AV and CV crashes. It is, therefore, here termed a crash selection bias, or more precisely, an unrecorded crash selection bias.

2.1.2.1. Adjustment for Unrecorded CV Crashes

Schoettle and Sivak (2015) [2.1] correctly attempted to adjust for the unrecorded CV crash selection bias identified in their original data reproduced in Table 2.1. The researchers adjusted for the fact that their CV crash estimate of 4.1 in Table 2.1 (note h) was biased low because of what they called "underreporting of crashes that do not involve a fatality." They increased the CV crash count by adding an estimate of unreported CV crashes to their original police-reported CV crashes of 5,687,057 (Table 2.1, note f). To estimate the unreported CV crashes, they used underreporting estimates of 60% for PDO crashes and 24% for injury crashes (Blincoe et al., 2015, Table 1-3 and Table 5-14) [2.5]. Schoettle and Sivak (2015) [2.1] then increased the number of CV crashes in Table 2.1 (note f) by multiplying the original police-recorded CV injury crash estimate by a correction factor of 1.32 to adjust for 24% unreported injury crashes. They also multiplied the original police-recorded PDO crash estimate by a correction factor of 2.5 to adjust for 60% unreported PDO crashes. Table 2.2 (note f) shows their estimated CV crash count increased to 12,295,177 after this adjustment.

Table 2.2 is the same as Table 2.1 except that Schoettle and Sivak (2015) [2.1] increased the number of CV crashes to 12,295,177 (note f) to include estimates of unreported injury and PDO crashes in their correct effort to reduce unrecorded CV crash selection bias. As a result, the upwardly revised CV crash count in Table 2.2 (note f) estimates a 2013 U.S. CV crash rate of 4.1 (note h), substantially more than their original CV crash rate estimate of 1.9 in Table 2.1 (note h).

Schoettle and Sivak (2015) [2.1] noted from their revised data in Table 2.2 that their AV crash rate estimate of 9.1 (Table 2.2, note d) was still substantially higher than the upwardly revised estimate of the 2013 U.S. CV crash rate of 4.1 (Table 2.2, note h). Thus, Schoettle and Sivak (2015) [2.1] concluded that the "current best estimate" is that AVs "have a higher crash rate per million miles traveled than conventional vehicles." Indeed, using the epidemiologic method in Table 1.1 in Chapter 1, Schoettle and Sivak's (2015) [2.1] crash rate data in Table 2.2 estimate an AV crash IRR of (9.1 ÷ 4.1) or 2.2.

However, these investigators cautioned that they could not rule out the reverse possibility that the "actual rates for self-driving vehicles are lower than for conventional vehicles," based on the "overlap" of the 95% confidence limits in Table 2.2 and applying a traditional statistical significance criterion.

TABLE 2.2 Schoettle and Sivak (2015): Revised CV crash data.

	AV[a]	2013 U.S. CV[e]
Total crashes	11[b]	12,295,177[f]
MVMT	1.214626[c]	**2,988,323**[g]
Crash rate	9.1[d]	4.1[h]
95% CI	(4.5-16.3)	(3.5-4.7)

Crashes, mileage, and crash rates for AVs and revised 2013 U.S. CVs.

A **bold** number in the 2×2 table above means no identified error or bias.

The underlined words in the notes below are revised in a later table.

Notes:

[a-e, g] Same as Table 2.1.

[f] The total CV crash count now includes an additional 24% of injury CV crashes, and an additional 60% of PDO CV crashes, based on estimates of unreported crashes in Blincoe et al. (2015) [2.5].

[h] 2013 U.S. CV police-recorded plus police-unrecorded crash rate estimate.

2.1.3. AV Site Selection Biases: Environment, City Streets (Table 2.2, Note d)

Schoettle and Sivak (2015) [2.1] correctly cautioned that if the AV testing had been conducted at a national level, the AV crash rate estimate might be even higher than 9.1 crashes per MVMT (Table 2.2, note d). They noted that the AVs they studied traveled only a few miles (1.2 MVMT) relative to the CVs (3 million MVMT), and the AV miles were in "less challenging" (e.g., non-snowy) road conditions. Therefore, they cautioned that the exposure of AVs to environmental conditions "has not yet been representative of the exposure for conventional vehicles." In other words, they imply that their estimate of 9.1 AV crashes per MVMT (Table 2.2, note d) would likely be even higher if the AV test sites (i.e., the AV Operational Design Domain, or ODD) included the entire United States, with a broader range of weather conditions. This plausible observation further supports their "current best estimate" that AVs have a higher crash rate than CVs.

In addition to this site selection bias from environmental conditions, there may have been a site selection bias from city street/highway differences. The Google AV ODD was mainly on city streets. But the U.S. national CV dataset also includes mileage on highways, interstates, and rural areas. Urban driving has a substantially higher crash rate than driving at all other sites, at all non-fatal crash Severity Levels (Blanco et al., 2016, pp. 27–28, Figures 7–9) [2.6].

Environmental and city street biases are just two of many types of site selection bias. Such biases arise when the site selected for the AV test has a different CV crash rate than the CV comparison site. One method to adjust the AV crash rate for this difference is standardization (Chapter 3, Section 3.4; Appendix H). However, site selection bias cannot be adjusted for AV tests at non-CA AV sites using the standardization method if the AV crash and mileage data specific to those sites have not been publically released by the AV developers. To date, AV crashes and overall AV mileage have been publicly released only in CA, where it is required by state law [2.7]. Even in CA, AV developers do not publicly release the specific AV mileage for each city within their ODD.

2.2. Critical Analysis of Schoettle and Sivak (2015) Study

Using the epidemiologic IRR method described in Sections 1.1 and 1.2, Schoettle and Sivak's (2015) [2.1] data in Table 2.2 estimate an AV crash IRR of 2.2 (exact mid-P 95% CI 1.16–3.83, P-value = 0.019). The AV crash IRR estimate of 2.2 indicates that AVs increase crashes by 120% compared to 2013 U.S. national CVs. Nonetheless, these data, if valid, are also reasonably compatible with this AV rate ranging from an increase of 16% to 283% versus 2013 U.S. CVs. If the data were without error or further bias, the IRR of 2.2 estimated in Table 2.2 from Schoettle and Sivak's (2015) [2.1] revised CV MVMT, therefore, appears to continue to support their conclusion that AVs "have a higher crash rate per million miles traveled than conventional vehicles."

However, multiple issues (a minor systematic error, biases, and best practice violations) remain in Schoettle and Sivak's (2015) [2.1] revised data in Table 2.2, even though Table 2.2 represents a revision by Schoettle and Sivak (2015) [2.1] of their original data in Table 2.1.

2.2.1. Minor Systematic Error: Wrong Percentage of Unrecorded CV Injury Crashes (Table 2.2, Note f)

Section 2.1.1 showed how Schoettle and Sivak (2015) [2.1] correctly attempted to eliminate the unrecorded crash selection bias in their original crash data in Table 2.1 by adding unreported CV crashes to the police-recorded CV crashes (Table 2.2, note f). However, Schoettle and Sivak (2015, p. 14) [2.1] used the erroneous estimate by Blincoe et al. (2015, p. 2) [2.5] that "24 percent of all injury crashes are not reported to the police." However, Blincoe et al. (2015) misstated their own results. Table 2.3 extracts the relevant rows from Table 1-3 in Blincoe et al. (2015, p. 13) [2.5].

Table 2.3 shows that the estimate of "24% injury crashes not reported"—cited by Blincoe et al. (2015, p. 2) [2.5] and Schoettle and Sivak (2015, p. 14) [2.1]—is the "Percent Unreported People in Injury Crashes." In other words, 24.1% is the percent of unreported injured victims, not the percent of unreported injury crashes. Thus, the 24.1% estimate is the number of "People in Injury Crashes" that were "Not Police-reported" (952,120) divided by the "Total" number of "People in Injury Crashes" (3,954,504).

The correct percent of unreported injury crashes is 39.7% (lower right cell in Table 2.3). This 39.7% estimate is the number of "Not Police-reported" "Injury Crashes" (1,178,391) divided by the "Total" "Injury Crashes" (2,969,963). Thus the 39.7% estimate is the percent unreported crashes in which there were one or more injured people.

In other words, Schoettle and Sivak (2015) [2.1] were misled by Blincoe et al. (2015, p. 2) [2.5] erroneous statement that "24 percent of all injury crashes are not reported to the police." Hence, Schoettle and Sivak (2015) [2.1] used the wrong level of analysis (at the victim level rather than the crash level) for their unreported injury crash estimate. The results of using the incorrect figure of 24.1% unreported CV injury crashes (rather than the correct 39.7%) were systematic downward errors in the total CV crash count estimate (Table 2.2, note f) and the CV crash rate estimate (Table 2.2, note h). Consequently, the AV crash IRR estimate of 2.2 from their revised Table 2.2 data is still biased high by unrecorded crash selection bias.

2.2.1.1. Correction of Minor Systematic Error: CV Crashes (Table 2.2, Note f)

The correct percentage of unreported injury crashes from Blincoe et al.'s (2015) [2.5] data replicated in Table 2.3 is 39.7%, not 24.1%, as noted above.

A revised analysis here corrects the Schoettle and Sivak (2015) [2.1] CV crash rate estimate by using the same method that Schoettle and Sivak (2015) [2.1] used to correct for unreported crashes but with the correct 39.7% unreported injury crash estimate (Table 2.3,

TABLE 2.3 Incidence summary: 2010 total police-reported and unreported injuries.

Severity	Police-reported	Not Police-reported	Total	Percent Unreported
		People in Injury Crashes		
Total Injuries	3,002,385	952,120	3,954,504	24.1%
		Crashes		
Injury	1,791,572	1,178,391	2,969,963	39.7%

Reproduction of portions of Table 1-3 in Blincoe et al. (2015, p. 13) [2.5].

lower rightmost cell). The correct multiplier factor for unreported injury crashes is then 1.66, not the 1.32 used by the UMTRI investigators. This correction increased the estimated CV crash count and rate, reducing the final AV crash IRR estimate.

This minor systematic error is corrected in a reanalysis of the Schoettle and Sivak (2015) [2.1] data in Section 2.3. However, there was a potential remaining unreported crash selection bias despite this minor systematic error correction. As a result, the estimated percentages of unreported injury crashes (39.7%) and unreported PDO crashes (60%) may still be too low, according to naturalistic driving study (NDS) data, where unrecorded crashes can be as high as 95%. (See Table E.1 in Appendix E for the range in the percentages of unreported and unrecorded crashes in various studies.)

2.2.2. Potential Site Selection Bias from Non-CA AV Tests (Table 2.2, Note c)

There are no legal requirements in any state other than CA for all AV crashes to be reported in a publicly available database. Thus, potential downward bias in the AV crash rate results from underreporting AV crashes while fully reporting the total AV mileage from Google, Delphi, and VW/Audi tests at non-CA sites. In other words, all 11 AV crashes in Table 2.2 (note b) were Google AV crashes in Mountain View and vicinity. However, the Google, Delphi, and VW/Audi AV MVMT in Table 2.2 (note c) included non-CA mileage. (Google also conducted AV tests in Austin, Texas [TX]; Delphi in multiple states; and VW/Audi in CA and Nevada.)

If non-CA AV crashes were underreported and non-CA AV mileages were voluntarily reported accurately, the inclusion of non-CA AV mileage in the AV MVMT tabulation would upwardly bias the AV MVMT (Table 2.2, note c). This bias is another type of site selection bias. This type of bias is caused by non-CA AV test sites with no requirements to report AV crash and mileage data to the state DMV (unlike CA). Again, the result of this site selection bias would be a downward bias in the AV crash rate estimate and the AV IRR estimate because of the non-CA test sites for prototype AVs.

2.2.2.1. Removal of Possible Site Selection Bias from Non-CA AV Tests

In short, there was a possible bias from AV crash underreporting in states other than CA because the other states' DMVs did not require such reporting. Therefore, the AV mileage from non-CA sites could bias the total AV crash rate too low. Hence, all AV mileage from Delphi, VW/Audi, and Google from states outside CA was removed in the reanalysis of the corrected Schoettle and Sivak (2015) [2.1] data as per Section 2.3.

2.2.3. Best Practice Violation: Combining Heterogeneous Companies (Table 2.2, Notes b and c)

Schoettle and Sivak (2015) [2.1] combined the heterogeneous AV crash and mileage data from Google, Delphi, and VW/Audi, a best practice violation of heterogeneity (Section 1.3). All 11 crashes were from Google AVs, so the Google AV crash IRR was nonzero. However, the Delphi Automotive and VW/Audi AV crash IRRs were zero since they had no AV crashes. Unless the strata are tested and shown to be homogeneous, the best practice is to analyze each stratum's data separately rather than combine the strata (Section 1.3).

2.2.3.1. Remove Best Practice Violation: Stratify Heterogeneous Companies

All 11 AV crashes from December 2009 to September 30, 2015 (Table 2.2, note b) were from Google. Therefore, the heterogeneous Delphi and VW/Audi data were eliminated in the reanalysis in Section 1.3. Only the Google AV crash and mileage data were reanalyzed. The Delphi and VW/Audi data were too sparse for separate analyses to be informative, so the data from those two companies were not examined further.

2.2.4. Best Practice Violation: Combining Heterogeneous Severities (Tables 2.1 and 2.2, Notes b and f)

Schoettle and Sivak (2015) [2.1] also combined data across crash severities (fatality, injury, and PDO) for both AVs and CVs (Tables 2.1 and 2.2, notes b and f). AV crash severity levels were tested as per Section 1.3 and found to be heterogeneous. CV crash severity levels are well known to be heterogeneous (Knipling, 2017 [2.8]; Knipling and Wåhlberg, 2019) [2.9]. This issue is another best practice violation of heterogeneity (combining data across crashes of different severity without first testing the strata to ensure they are homogeneous).

2.2.4.1. Remove Best Practice Violation: Stratify Heterogeneous Crash Severities

The crash heterogeneity best practice violation was avoided in the current reanalysis by separately analyzing (i.e., stratifying) the AV and CV crash counts in Table 2.2 (notes b and f) by crash severity level.

2.3. Reanalysis of Corrected Schoettle and Sivak (2015) Data

Table 2.4 reanalyzes the Schoettle and Sivak (2015) [2.1] data in Table 2.2, after correcting the systematic errors, adjusting the known biases, and removing the best practice violations noted in Sections 2.1 and 2.2.

As per Table 2.2, Table 2.4 compares the AV crash rates (now restricted to Google AVs) from 2009 to September 30, 2015, to the 2013 U.S. national CV crash rates (NHTSA, 2015) [2.3]. However, Table 2.4 corrects the errors, biases, and best practice violations in the Table 2.2 data as identified in the Section 2.2 analysis of the Schoettle and Sivak (2015) [2.1] study. Table 2.4 also stratifies the data by crash severity.

Table 2.4A data estimate a Google AV fatal crash IRR of zero, indicating that Google AVs decrease the fatal crash rate by 100% compared to the 2013 U.S. national CV fatal crash rate [2.3]. Nonetheless, these data, if valid, are also reasonably compatible with this Google AV fatal crash rate ranging from a 100% decrease to a 28,440% increase versus the 2013 U.S. national CV fatal crash rate.

Table 2.4B data estimate a Google AV injury IRR of 2.2, indicating that Google AVs increase the injury crash rate by 117% compared to the U.S. national CV injury crash rate. Nonetheless, these data, if valid, are also reasonably compatible with this Google AV injury

TABLE 2.4 Schoettle and Sivak (2015): Corrected data stratified by crash severity.

	Google AV[a]	2013 U.S. CV[e]	Google AV[a]	2013 U.S. CV[e]
	A. Fatal		**B. Injury**	
Crashes	**0[b]**	**30,057[f]**	2[b]	2,637,467[f]
MVMT	**1.0437[c]**	**2,988,323[g]**	**1.0437[c]**	**2,988,323[g]**
Crash rate	0[d]	0.0101[h]	1.92[d]	0.883[h]
IRR (95% CI)[i]	0.0 (0–285.4), P = 0.9896		2.2 (0.36–7.2), P = 0.3017	
	C. PDO		**D. Total crashes**	
Crashes	9[b]	10,095,037[f]	11[b]	12,762,561[f]
MVMT	**1.0437[c]**	**2,988,323[g]**	**1.0437[c]**	**2,988,323[g]**
Crash rate	8.62[d]	3.38[h]	10.54[d]	4.27[h]
IRR (95% CI)[i]	2.6 (1.2–4.7), P = 0.0138		2.5 (1.3–4.3), P = 0.0085	

A **bold** number in a 2×2 subtable above indicates no known error or bias.

The underlined words in the notes below are revised in later tables.

Notes:

[a] Same as Table 2.1, except Google data only (Delphi and VW/Audi data were removed).

[b] Same as Table 2.1, except Google AV crashes are stratified by severity level.

[c] Same as Table 2.1, except Google AV CA mileage only (Google AV non-CA mileage was removed).

[d-e, g] Same as Table 2.1.

[f] Same as Table 2.2, except CV crashes are stratified by severity level. Subtables B and D include 39.7% unrecorded injury CV crashes instead of 24%, and subtables C and D include 60% unrecorded PDO CV crashes (see Section 2.2.1).

[h] Same as Table 2.2.

[i] The IRR, exact mid-P confidence levels, and two-tailed P were estimated with the StatsDirect rate ratio program [2.10].

crash rate ranging from a 64% decrease to a 617% increase versus the 2013 U.S. national CV injury crash rate.

Table 2.4C data estimate a Google AV PDO IRR of 2.6, indicating that Google AVs increase the PDO crash rate by 155% compared to U.S. national CVs. Nonetheless, these data, if valid, are also reasonably compatible with this Google AV PDO crash rate ranging from a 25% to 368% increase versus the 2013 U.S. national CV PDO crash rate.

The Google AV fatality, injury, and PDO IRR estimates in Tables 2.4A–C are homogeneous (P = 0.17). Therefore, the AV and CV crashes can be validly combined across severities (Table 2.4D).

Table 2.4D data estimate a Google AV total crash IRR of 2.5, a slight increase from the 2.2 IRR estimate derived from the Table 2.2 data. This Google AV total crash IRR estimate of 2.5 indicates that Google AVs increase the total crash rate by 147% compared to the U.S. CV total crash rate. Nonetheless, these data, if valid, are also reasonably compatible with the Google AV total crash rate ranging from a 30% to 330% increase versus the 2013 U.S. CV total crash rate. The revised Table 2.4D data, therefore, appear to continue to support Schoettle and Sivak's (2015) [2.1] original main conclusion that AVs "have a higher crash rate per million miles traveled than conventional vehicles." However, the data in Table 2.4, even though considerably corrected and adjusted from the original Schoettle and Sivak (2015) [2.1] data in Table 2.2, may still have a remaining bias as shown in the following Sections.

2.4. New Methods to Remove Unrecorded Crash Selection Bias and Site Selection Bias

Schoettle and Sivak (2015) [2.1] correctly recognized (Section 2.1.2) that an estimate of unrecorded CV crashes had to be added to the police-recorded CV crashes to equalize the criteria for selecting AV and CV crashes (i.e., all crashes, recorded and unrecorded). They thereby avoided biasing the AV IRR estimate upwards from unrecorded crash selection bias caused by mistakenly selecting all AV crashes, but only police-recorded CV crashes.

However, the selection of unreported CV crash percentages from studies in Appendix E, Table E.1 is arbitrary. For example, Section 3.5 references a study showing that selecting different percentages of unreported or unrecorded crashes from various published survey studies (see Appendix E) can make AV crash rates appear better, equal, or worse than CV crash rates.

The percentage of unreported injury CV crashes that Schoettle and Sivak (2015) [2.1] used to estimate the total CV crashes in Table 2.2 (note f) was corrected in Tables 2.4B–D. However, these percentages of unreported crashes were still based solely on telephone survey data from Blincoe et al. (2015) [2.5]. There are, in general, limitations of using any survey data to estimate unreported and unrecorded crash percentages (see Appendix E).

Hence, two new analysis methods were developed in this study to avoid this arbitrary selection of unreported and unrecorded CV crashes. These methods were intended to remove any unreported and unrecorded crash selection bias, as well as site selection bias, in CV crash count estimates (Table 2.4, note f).

2.4.1. First New Method to Remove Unrecorded Crash Selection Bias: Use Only Police-Recorded AV Crashes

This new correction method compares only police-recorded AV crashes (rather than all AV crashes) to police-recorded CV crashes. This method removes unrecorded crash selection bias because the criterion for crash selection is the same for AV and CV crashes (i.e., AV and CV crashes must have a police record to be counted). As a result, it is no longer necessary to estimate an uncertain number of unrecorded CV crashes. Therefore, an exhaustive search was done for the police-recorded crash that matched each of the 11 Google AV crashes investigated by Schoettle and Sivak (2015) [2.1]. (The matching method was described in Section 1.4.)

Table 2.5 presents the results, stratified by crash severity level to avoid heterogeneity, for the Google AV crashes that could be successfully matched with a police record (Table 2.5, note b). These are compared to the 2013 U.S. national CV crashes with a police record (NHTSA, 2015) [2.3].

Table 2.5 compares, stratified by crash severity level, the police-recorded crash rate of the matched AV crashes (Table 2.5, note d) to the police-recorded 2013 U.S. national CV crash rate (NHTSA, 2015) [2.3] (Table 2.5, note h). Unrecorded crash selection bias is avoided by having the identical selection criterion (police-recorded crashes only) for the AV-exposed and AV-unexposed (i.e., CV) fleets.

TABLE 2.5 Method to remove unreported and unrecorded crash selection bias from the Schoettle and Sivak (2015) data: Use only police-recorded AV crashes.

	Google AV[a]	2013 U.S. CV[e]	Google AV[a]	2013 U.S. CV[e]
	A. Fatal		**B. Injury**	
PR* crashes	**0**[b]	**30,057**[f]	**0**[b]	**1,591,000**[f]
MVMT	**1.0437**[c]	**2,988,323**[g]	**1.0437**[c]	**2,988,323**[g]
Crash rate	0.0[d]	0.0101[h]	0.0[d]	0.53[h]
IRR (95% CI)[i]	0.0 (0–285.4), P = 0.9896		0.0 (0–5.39), P = 0.5737	
	C. PDO		**D. Total crashes**	
PR* crashes	**0**[b]	**4,066,000**[f]	**0**[b]	**5,687,057**[f]
MVMT	**1.0437**[c]	**2,988,323**[g]	**1.0437**[c]	**2,988,323**[g]
Crash rate	0.0[d]	1.36[h]	0.0[d]	1.89[h]
IRR (95% CI)[i]	0.0 (0–2.11), P = 0.2417		0.0 (0–1.51), P = 0.1372	

* Police-recorded AV and CV crashes only, stratified by crash severity.

A **bold** number in a 2×2 subtable above indicates no identified error or bias.

Notes:

a, c–e, g–i Same as Table 2.4.

b AV crashes with a matching police record in the CA SWITRS [2.11] database.

f Estimated CV crashes with a police record in 2013, stratified by crash severity (NHTSA, 2015, p. 2) [2.3].

Cells with note "b" in Tables 2.5A–D represent the Google AV crashes with a matching police record. As stated, police records were matched using the verification method in Section 1.4 with the SWITRS database [2.11] for the Schoettle and Sivak (2015) [2.1] study period from 2009 to September 30, 2015. As a result of this matching method, none of the 11 Google AV crashes (Table 2.4B, note b) investigated by Schoettle and Sivak (2015) matched any crash in the SWITRS [2.11] police record database. Therefore, the main difference between Tables 2.4 and 2.5 is that the AV crashes become zero in Tables 2.5B–D (note b) compared to Tables 2.4B–D (note b).

Another substantial difference between Tables 2.4 and 2.5 is that the CV crashes in Table 2.5 (note f) are now police-recorded crashes only. In contrast, in Table 2.4 (note f), the CV crashes are police-recorded plus an arbitrary selection of an unknown number of unrecorded crashes (Appendix E, Table E.1).

Note that the MVMT data in Table 2.5 (notes c and g) are the same as the MVMT data in the previous Table 2.4. Also, the fatality data in Table 2.5A are the same as in Table 2.4A because all fatalities are police-recorded.

Table 2.5A data estimate a Google AV known police-recorded injury IRR of zero. Table 2.5A has the same AV and CV data, IRR, and CI estimates as per Table 2.4A.

Table 2.5B data estimate a Google AV known police-recorded injury IRR of zero, indicating that Google AVs decreased the police-recorded injury crash rate by 100% compared to the 2013 U.S. national CV crash rate [2.3]. Nonetheless, these data, if valid, are also reasonably compatible with this Google AV injury crash rate ranging from that 100% decrease to a 439% increase versus the 2013 U.S. police-recorded CV injury crash rate [2.3].

Table 2.5C data estimate a Google AV known police-recorded Major PDO IRR of zero, indicating that Google AVs decrease the police-recorded Major PDO crash rate by 100% compared to the 2013 U.S. CV Major PDO crash rate [2.3]. Nonetheless, these data, if valid, are also reasonably compatible with this Google AV Major PDO crash rate ranging from that 100% decrease to a 111% increase versus the 2013 U.S. police-recorded CV Major PDO crash rate.

The Google AV fatality, injury, and PDO IRRs in Tables 2.5A–C are all zero, so the subtables are homogeneous ($P = 1$). Therefore, the crashes in Tables 2.5A–C (notes b and f) can be combined (Table 2.5D).

Table 2.5D data estimate a Google AV known police-recorded total crash IRR of zero, indicating that Google AVs decrease the total police-recorded crash rate by 100% versus the corresponding 2013 U.S. police-recorded CV crash rate. Nonetheless, these data, if valid, are also reasonably compatible with this Google AV total police-recorded crash rate ranging from a 100% decrease to a 52% increase versus the 2013 U.S. CV police-recorded total crash rate.

Note that the zero police-recorded crashes in Table 2.5 (note b) means that all 11 Google AV crashes in this study period were not recorded by police. That is, 100% of the AV crashes in Table 2.5 (note b) were unrecorded by police. That lack of police records for AV crashes was consistent with the 95% unrecorded CV crashes in the Strategic Highway Research Program (SHRP2) study (Blanco et al., 2016) [2.6] (see also Table E.1, last line, SHRP2). Thus, at a 95% unrecorded rate (1 in 20 crashes), it is not too surprising that none of the 11 AV crashes were recorded.

Nonetheless, the finding that none of the 11 AV crashes were recorded was still unexpected. Two of the 11 AV crashes represented in Table 2.4B (note b) were injury crashes. Injury crashes more often have a police record than PDO crashes (see Appendix E). For crash descriptions of the injury crashes, see Appendix E.4.1 for the July 1, 2015 injury crash and Appendix E.4.2 and Appendix C for the August 20, 2015 injury crash, which occurred during transition. It was also somewhat unexpected that none of the nine PDO AV crashes in Table 2.4C (note b) were recorded. At least two of the Major PDO crashes were likely above the CA repair cost threshold of $750 at the time, requiring at least a police report to be filed (although a police record in the state database might still not be entered by the police). (See Section 5.1.2.1 for the Major PDO crash description on March 2013, with an unspecified day, and the one on February 26, 2015.) None of the Minor PDO AV crashes had a matching police record, as expected.

This failure to file a police record for any of the AV crashes happened in some cases even though the police had been called to the crash scene according to the crash description in the Google Reports [2.2]. The local police must have decided not to file a formal Police Accident Report or enter the crash record in the state police database. Or an AV occupant or a driver of another vehicle may have reported the crash to the police. A local police department might write up the information in a brief bench report but choose not to file a formal Police Accident Report. It is evident that local police department decisions not to enter an official crash record in the state database also occur for CV crashes. However, more AV crashes than CV crashes may be unrecorded by the police for possible reasons discussed in Appendix E.5.

Table 2.5 was hence successful in adjusting for unrecorded crash selection bias. However, there was still a potential site selection bias in Table 2.5. The reason is that the Google AV fleet was tested in Mountain View and vicinity, where crash rates are not representative of the United States nationally (see Appendix K, Table K.1). The result is a site selection bias. See Section 2.1.3 for a discussion of site selection bias and Section 2.4.2 for a method to adjust for it and unrecorded crash selection bias.

2.4.2. Second New Method to Remove Unrecorded Crash Selection Bias and AV Site Selection Bias: AV Trend Analysis

Trend analysis is another new method (in addition to that in Section 2.4.1) for AV crash safety analyses to avoid the arbitrary selection of percentages of unrecorded injury and

PDO crashes from studies such as those shown in Table E.1. This new trend analysis method simply compares the AV crash rate data to itself. AV trend analysis has several advantages over trying to compare AV crash rates to CV crash rates. First, AV trend analysis avoids unrecorded crash selection bias from selecting arbitrary percentages of injury and PDO unreported or unrecorded crashes (Section 2.1.2). Second, AV trend analysis avoids site selection bias from different CV crash rates at different sites (e.g., Mountain View vs. U.S. national).

However, as shown in Section 2.4.2.2, AV trend analysis does not alleviate the need to compare the AV trend to the CV trend at the same site. For example, an upward trend in the AV crash rate does not necessarily mean that the ADS software updates worsen the AV crash rate. Conversely, a downward trend in the AV crash rate does not necessarily mean that the ADS software updates improve the AV crash rate. Therefore, a method is needed to test whether a trend in the AV crash rate arises from (1) technical changes made to the ADS software or hardware or (2) a general trend in the crash probability at an AV test site. This new AV trend analysis method was developed to test that question.

This new AV trend analysis method does not alleviate the need to remove site selection bias caused by CV crash rate differences between the AV site and the CV site when estimating an AV crash IRR. For example, if the <u>CV</u> crash rate at the AV site were low compared to the CV crash rate at the comparison site, it would falsely appear as if the <u>AV</u> crash rate were also low compared to the CV crash rate at the comparison site. In other words, the final IRR estimates in Table 2.5 for the Schoettle and Sivak (2015) reanalysis still have a potential remaining site selection bias. The trend analysis may avoid it, but Mountain View may still have a lower police-recorded CV crash rate than nationally. For example, residents of Mountain View who are in a crash may not report the crash to police as often as general residents in the United States. Or the Mountain View police department may not enter a formal police record in the state crash database as often as the average police department or highway patrol in the U.S. (see Appendix E regarding unrecorded crashes). Therefore, Table 2.5 does not have all site selection bias removed. It is, therefore, possible that other methods to reduce site selection bias, such as site standardization using CV crash rates, are needed to remove site selection bias from AV crash IRR estimates.

2.4.2.1. Trend Analysis of Google AV Cumulative Crash Rate

Figure 2.1 plots the cumulative crash rate by date of the 11 Google AV crashes analyzed by Schoettle and Sivak (2015) [2.1] for their study period from 2012 to September 30, 2015.

Figure 2.1 shows by visual inspection that Google AV crashes in Mountain View, CA, and vicinity (Google Reports [2.2]) were relatively sparse from 2012 to 2014. Therefore, the cumulative Google AV crash rate was primarily flat until February 26, 2015.

Is this 2012–2014 sparseness because Google did not voluntarily release some of its crashes in its Google Reports [2.2]? CA DMV regulations [2.7] require reporting AV crashes and AV mileage on public roads in CA. Still, these rules did not go into effect until September 16, 2014, making it more plausible that Google may not have publicly disclosed all its AV crashes on public roads until it was legally required. If Google did not count some crashes but did count its total AV mileage in these early periods, then these early AV crash rates would be biased low. However, there were also no crashes after the required reporting began on September 16, 2014, until February 26, 2015, more than five months later, suggesting the early crash sparseness may be valid.

After the flat stretch in the Google AV cumulative crash rate from October 2012 to February 2015, Figure 2.1 shows by visual inspection that the Google AV cumulative crash rate exhibits an upward trend from the Major PDO crash on February 26, 2015, to the injury

FIGURE 2.1 Google AV cumulative crash rate trend in the entire Schoettle and Sivak (2015) [2.1] study period. X-Axis: Month and year from October 2012 to October 2015. **Y-Axis**: Google AV cumulative crash rate (cumulative AV crashes ÷ cumulative AV MVMT). **Symbols**: Eleven crashes classified as autonomous. **Filled green circles:** Nine crashes during autonomous mode. **Open circles:** Two crashes during the transition from autonomous to manual mode (2/26/15, 8/20/15). **Large circles:** Level 1 Severity (two Severe crashes: autonomous 7/1/15, transition 8/20/15). **Medium circles:** Level 2 Severity (two Major PDO crashes: autonomous 3/-/13, transition 2/26/15). **Small green circles:** Level 3 Severity (seven Minor PDO crashes).

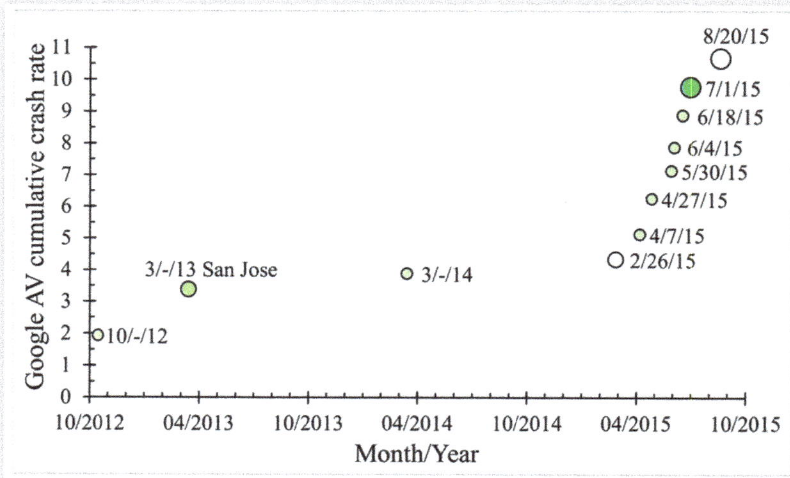

crash on August 20, 2015, the last crash before the end date of the Schoettle and Sivak (2015) [2.1] study period on September 30, 2015. Four hypotheses to explain this upward trend in 2015 are presented in the next section.

2.4.2.2. Hypotheses for Upward Trend in 2015 Google AV Cumulative Crash Rate

2.4.2.2.1. Hypothesis 1. Changes in Google Testing Protocols. The first hypothesis is that there were changes in the Google testing protocols in 2015. The increase in Google AV crash rates in 2015 was also recognized in a later study by Teoh and Kidd (2017) [2.12], reviewed in Chapter 5. They hypothesized (Section 5.2.1) that the 2015 crash rate increase was from "increased usage on higher speed roads" or "more aggressive testing practices." Unfortunately, Teoh and Kidd (2017) [2.12] presented no data or evidence for this hypothesis, so it will not be further investigated.

2.4.2.2.2. Hypothesis 2. Negative Learning by the Google Neural Network. The second hypothesis is that the upward Google AV crash rate trend in 2015 results from negative learning by the neural network underlying the Google Automated Driving System (ADS). Google states that it frequently updates its onboard ADS software by putting the on-road test results through its simulator and trying multiple variations on the crash scenarios. The "improved" software is then downloaded to the vehicle with the expected outcome of reducing crashes and disengagements by the test drivers' and ADS. But an ADS neural network might have a negative learning effect, where more training might cause more driving performance errors, not fewer. This hypothesis is discussed further in Appendix M.

2.4.2.2.3. Hypothesis 3. Upward Trend in CV Cumulative Crash Rate. The third hypothesis is that the upward trend in the 2015 Google AV cumulative crash rate seen in Figure 2.1 simply reflects an upward trend in the probability of a vehicle crash in Mountain View in 2015, for whatever reason. If so, then hypotheses 1 and 2 are not needed to explain the upward trend in the AV crash rate.

Hence, the CV cumulative crash rate trend in Mountain View was estimated to investigate and test this third hypothesis. First, all Mountain View police-recorded CV crashes on city streets were pulled from the SWITRS database [2.11] from March to September 2015. Next, the total CV crashes were divided by the Mountain View City street mileage, prorated per month from the Highway Performance Monitoring System (HPMS) database [2.13]. The police-recorded CV cumulative crash rate per month was then plotted, and its trend compared to the upward AV crash rate trend over the same period.

2.4.2.2.4. Hypothesis 4. Downward Trend in Test Drivers' Cumulative Disengagement Rate. The focus of this book is on crashes, not disengagements. However, if disengagements affect crashes, such an interaction effect is within scope.

Consequently, the fourth hypothesis is that the increase in AV crash rate in 2015 was caused by a reduction by Google AV test drivers of their voluntary disengagement rate in 2015. For example, suppose a test driver believes a crash is imminent, and manual control is necessary to avoid the crash. In that case, the test protocols call for the test driver to disengage the ADS manually. Conversely, the test driver might choose to reduce their voluntary disengagement rate for some reason. The imminent crash would, therefore, not be avoided by assuming manual control. In such a case, a lower test driver disengagement rate predicts a higher AV crash rate.

This test driver disengagement hypothesis can be tested empirically with existing data. It is simply necessary to compare the trend in the AV test drivers' cumulative disengagement rate to the trend in the AV cumulative crash rate over the same period.

Test driver disengagement is only one of the three types of disengagements from autonomous mode. The others are planned disengagements for testing purposes and disengagements by the ADS. Planned disengagements for testing purposes are not predicted to interact with crash rates and are not addressed further. The relationship between ADS disengagements and crashes is tested in Section 5.3.6 in Chapter 5.

2.4.2.3. Tests of Hypotheses 3 and 4 for Upward Trend in 2015 Google AV Cumulative Crash Rate

The data in Figure 2.2 are used to tests hypotheses 3 and 4. Figure 2.2 plots the cumulative crash rates of the last eight AV crashes in Figure 2.1 for 2015 and compares them to the Mountain View CV cumulative crash rate and the test driver cumulative disengagement rate over the same period from February 1, 2015, to August 31, 2015, on the X-axis.

Figure 2.2A shows the Google AV cumulative crash rate in Mountain View and vicinity at eight Google AV crash dates. The linear regression fit shows an upward trend ($r = 0.977$, df = 6, $P = 0.00003$). Figure 2.2B shows the Mountain View police-recorded CV cumulative crash rate at month-end. The linear regression fit shows an upward trend ($r = 0.832$, df = 5, $P = 0.02$). Figure 2.2C shows the Google AV test drivers' cumulative disengagement rate at month-end. The linear regression fit shows a downward trend ($r = -0.932$, df = 5, $P = 0.002$).

Hypothesis 3 was tested by comparing the cumulative crash rate for Google AVs (Panel A) to Mountain View CVs (Panel B) over the same period. Figure 2.2B shows by visual inspection that the upward trend in the Google AV cumulative crash rate in Mountain View and vicinity (Figure 2.2A) is associated with an upward trend in the Mountain View

FIGURE 2.2 Comparison of trends in the last eight months of the Schoettle and Sivak (2015) [2.1] study period in 2015. X-Axis: Month from February 2015 to September 2015. **A. Y-Axis**: Google AV cumulative crash rate. Mountain View and vicinity. **Symbols**: Eight Google AV crashes at dates shown; see Figure 2.1 caption for symbol key. **B. Y-Axis**: Police-recorded CV cumulative crash rate. Mountain View, CA, at month-end. **C. Y-Axis**: Google AV test drivers' cumulative disengagement rate (cumulative disengagements ÷ cumulative MVMT). Mountain View, CA, at month-end. The dotted line in each panel is the linear regression fit.

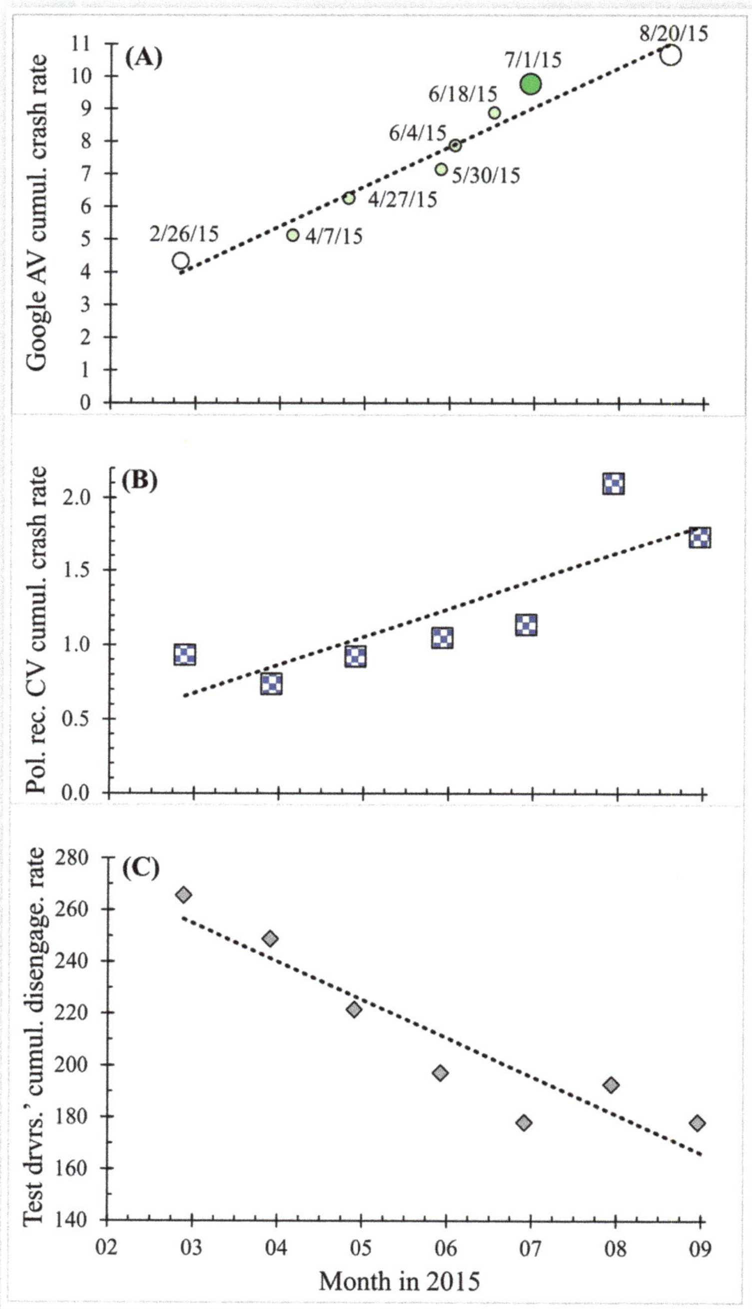

CV cumulative crash rate (Figure 2.2B). After interpolating the CV cumulative crash rate at the eight Google AV crash dates, the correlation coefficient between panel A and panel B is $r = 0.80$ (df = 6, $p = 0.017$). This comparison suggests that the upward trend in the Google AV cumulative crash rate is biased by a general upward trend in the general probability of a crash in Mountain View in 2015. The influence could not have been in the reverse direction (AV crashes increasing the crash counts in police records) because none of the AV crashes were police-recorded. In short, Figure 2.2B confirms Hypothesis 3.

As an aside, note that the Google AV cumulative crash rate (Figure 2.2A) is about eight times higher on an absolute scale than the police-recorded cumulative CV crash rate (Figure 2.2B). The reason is unrecorded crashes. All crashes (police-recorded and not police-recorded) are captured by Google AVs and plotted in Figure 2.2A. However, only police-recorded crashes are plotted in Figure 2.2B. However, this absolute difference in crash rates is precisely the reason why a trend analysis is beneficial. Unrecorded crash selection bias and site selection bias do not bias a trend analysis. The absolute AV and CV crash rates are irrelevant to the main finding that the AV and CV cumulative crash rate estimates have a similar upward trend at the same site in the same period.

Hypothesis 4 was tested by comparing the cumulative crash rate estimate for Google AVs (Panel A) to the Google AV test drivers' cumulative disengagement rate estimate (Panel C). Figure 2.2C confirms Hypothesis 4. Namely, the upward trend in the Google AV cumulative crash rate estimate for 2015 in Mountain View and vicinity (Figure 2.2A) is associated with a downward trend in the test drivers' cumulative disengagement rate estimate (Figure 2.2C) at the same site in the same period. After interpolating the test drivers' cumulative disengagement rate estimate at the Google AV crash dates, the correlation coefficient between the data in Panel A and Panel C is $r = 0.86$ (df = 6 and $p = 0.001$). Therefore, the data confirm that the downward trend in the test drivers' cumulative disengagement rate estimate could have contributed to the upward trend in the Google AV cumulative crash rate estimate.

Note that this confirmation of Hypotheses 3 and 4 does not disconfirm Hypotheses 1 and 2 but makes them unnecessary to explain the upward trend in the Google AV cumulative crash rate estimate in Figure 2.2A.

2.5. Summary of Reanalysis of Schoettle and Sivak (2015) Data

Section 2.1.1 reviewed the Google AV CA test data from 2009 to September 30, 2015, analyzed by Schoettle and Sivak (2015) [2.1]. Section 2.2 corrected errors and biases. Section 2.3 reanalyzed the updated data, and Section 2.4 proposed two new methods to remove any remaining biases. Several findings emerged.

The corrected Schoettle and Sivak (2015) [2.1] data in Table 2.4 in Section 2.3 appear at first to continue to confirm Schoettle and Sivak's (2015) [2.1] "current best estimate" that AVs "have a higher crash rate per million miles traveled than conventional vehicles." Indeed Tables 2.4B–D show that the Google AV crash rate was higher than the 2013 U.S. national CV non-fatal crash rate for injury, PDO, and total AV crashes, respectively.

However, these higher AV versus CV non-fatal crash rate estimates in Tables 2.4B and C were dependent upon the particular percentage of unrecorded CV crashes selected by Schoettle and Sivak (2015) [2.1] (Section 2.1.2 and Appendix E, Table E.1).

Therefore, two new methods were developed to eliminate any remaining possibility of unrecorded crash selection bias. In the first method (Section 2.4.1), the analysis was restricted to police-recorded crashes for AVs and CVs. As a result, the AV police-recorded crash IRR declined from 2.5 (Table 2.4D) to zero (Table 2.5D), a 100% decline. This new result with Schoettle and Sivak's (2015) [2.1] crash data no longer supports their "current best estimate" that Google AVs have a higher crash rate than CVs.

In the second method (Section 2.4.2), a new trend analysis method eliminated any possibility of unrecorded crash selection bias and reduced site selection bias. Figure 2.2A (Section 2.4.2.2) found an upward trend in the 2015 Google AV cumulative crash rate estimate, suggesting negative learning by the Google AV neural networks or other explanations (e.g., see Section 5.2.1). However, the Mountain View CV cumulative crash rate estimate also increased over this same period (Figure 2.2B). Also, the Google test drivers disengaged from autonomous mode less frequently over this same period. It follows that the upward trend in the Google AV cumulative crash rate in 2015 may be simply explained as a general increase in the crash probability in Mountain View during this period, a tendency for fewer test driver disengagements, or both.

However, these new findings from the trend analysis (Figure 2.2) imply that the Google ADS neural networks did not have positive learning. If there were positive learning, the upward trend in the AV cumulative crash rate in Figure 2.2A would have been less steep (or even downward) compared to the upward trend in the CV cumulative crash rate in Figure 2.2B. The downward trend in the test drivers' cumulative disengagement rate in Figure 2.2C suggests some positive neural network learning was occurring. Still, such neural network learning was not strong enough to lead to the expected improvement in Google AV crash rates (Figure 2.2A) compared to CV crash rates (Figure 2.2B). In other words, Figure 2.2 suggests that the Google AV cumulative crash rate in 2015 simply tracked the local CV cumulative crash rate, as well as the downward trend in test driver disengagements. In sum, the results in Figure 2.2 imply that the probability of an AV crash during the Google AV testing in 2015 simply reflected the local CV crash probability or lowered voluntary disengagements or both. There was no evidence of a Google ADS learning effect (either positive or negative) during this Google AV test period.

References

2.1. Schoettle, B. and Sivak, M., "A Preliminary Analysis of Real-World Crashes Involving Self-Driving Vehicles," Technical Report No. UMTRI-2015-34, The University of Michigan Transportation Research Institute, Ann Arbor, MI, 2015, http://www.umich.edu/~umtriswt/PDF/UMTRI-2015-34.pdf, accessed May 19, 2021.

2.2. Google Reports, "Google Self-Driving Car Project Monthly Reports," Mountain View, CA, May 2015–November 2016. Note: Monthly reports are no longer publicly available from Google. However, 19 monthly Google Reports from May 2015 to November 2016 are available at https://github.com/wgetsnaps/ca-dmv-autonomous-vehicle-accidents/tree/master/google-pdfs, accessed May 19, 2021.

2.3. NHTSA, "Traffic Safety Facts: 2013 Data," National Highway Traffic Safety Administration, DOT HS 812 169, Washington, DC, 2015, https://crashstats.nhtsa.dot.gov/Api/Public/ViewPublication/812169, accessed May 19, 2021.

2.4. Porta, M., *A Dictionary of Epidemiology*, 6th ed. (New York: Oxford University Press, 2014), 343 pp., http://irea.ir/files/site1/pages/dictionary.pdf, accessed May 19, 2021.

2.5. Blincoe, L., Miller, T.R., Zaloshnja, E., and Lawrence, B.A., "The Economic and Societal Impact of Motor Vehicle Crashes, 2010 (Revised)," Report No. DOT HS 812013, National Highway Traffic Safety Administration, Washington, DC, May 2015, https://trid.trb.org/view/1311862, accessed May 23, 2021.

2.6. Blanco, M., Atwood, J., Russell, S., Trimble, T. et al., "Automated Vehicle Crash Rate Comparison Using Naturalistic Data," Virginia Tech Transportation Institute, January 2016, https://vtechworks.lib.vt.edu/handle/10919/64420, accessed May 19, 2021.

2.7. California DMV, "Code of Regulations Title 13, Division 1, Chapter 1, Article 3.7—Testing of Autonomous Vehicles, §227.50," 2019, https://www.dmv.ca.gov/portal/uploads/2020/06/Adopted-Regulatory-Text-2019.pdf, accessed May 19, 2021.

2.8. Knipling, R.R., "Crash Heterogeneity: Implications for Naturalistic Driving Studies and for Understanding Crash Risks," *Transportation Research Record* 2663, no. 1 (2017): 117–125, doi:10.3141/2663-15, https://journals.sagepub.com/doi/abs/10.3141/2663-15, accessed May 19, 2021.

2.9. Knipling, R.R. and Wåhlberg, A.E., "The Heterogeneity Principle," in *Proceedings of the Tenth International Driving Symposium on Human Factors in Driver Assessment, Training and Vehicle Design*, University of Iowa Public Policy Center, Santa Fe, NM, 2019, 224–230, doi:10.3141/2663-15, https://drivingassessment.uiowa.edu/sites/drivingassessment.uiowa.edu/files/da2019_35_knipling_final_0.pdf, accessed May 19, 2021.

2.10. StatsDirect, Statistical Software, Version 3.2.8, September 16, 2019, https://www.statsdirect.com/, accessed September 22, 2021.

2.11. SWITRS, "California Highway Patrol Statewide Integrated Traffic Records System," https://iswitrs.chp.ca.gov/Reports/jsp/index.jsp, accessed May 19, 2021.

2.12. Teoh, E.R. and Kidd, D.G., "Rage Against the Machine? Google's Self-Driving Cars versus Human Drivers," *Journal of Safety Research* 63 (2017): 57–60, https://www.researchgate.net/profile/Eric_Teoh/publication/317265345, accessed May 19, 2021.

2.13. HPMS, "California Department of Transportation Highway Performance Monitoring System Annual Reports 2009–2014," https://dot.ca.gov/programs/research-innovation-system-information/highway-performance-monitoring-system, accessed May 19, 2021.

3

Google AV Crash Data versus Naturalistic Crash Data: Blanco et al. (2016)

Contents

Note: Refer to Appendix B on page 149 for an explanation of the abbreviations and the definitions of unique terms and phrases.

Blanco et al. (2016) [3.1], researchers at Virginia Tech Transportation Institute (VTTI), conducted the second published scientific study of AV crashes.

3.1. Summary of Blanco et al. (2016): Google AV versus SHRP2 CV

The Blanco et al. (2016) [3.1] research study, done under contract with Google, investigated the 11 Google AV crashes in CA during the test period from 2009 to October 31, 2015. (These were the same 11 Google crashes investigated by Schoettle and Sivak [3.2], whose data and analysis were reproduced and reanalyzed in Chapter 2.)

3.1.1. Google AV Original Crash Data

Google gave the VTTI team access to the crash videos and other vehicle data collected by the heavily instrumented Google AVs. Consistent with Schoettle and Sivak (2015) [3.2], Blanco et al. (2016, p. 8) [3.1] classified crashes that occurred during the transition from autonomous to manual mode in the same group as "autonomous." (See Appendix C for transition crash justifications, crash descriptions, and comments.) Blanco et al. (2016) [3.1] classified the 11 Google AV crashes into the same three crash Severity Levels used in the Second Strategic Highway Research Program (SHRP2) naturalistic driving study (NDS). (The SHRP2 NDS is hereafter abbreviated as SHRP2.) Blanco et al. (2016) [3.1] used the identical SHRP2 criteria as given in Appendix G to classify the severity of the Google AV crashes. The SHRP2 study is described in the following Section 3.1.2.

3.1.2. SHRP2 CV Original Crash Data

The SHRP2 CVs were fully instrumented (as were the Google AVs) so that all crash events were automatically recorded with video, radar, and inertial sensors. Therefore, no assumptions about unrecorded crashes had to be made to compare AV crash rates with SHRP2 CV crash rates, unlike Schoettle and Sivak (2015) [3.2] were required to assume to adjust for unrecorded crash selection bias (see Section 2.1.2).

The SHRP2 research study [3.1] involved 3,500 volunteer passenger-vehicle drivers, ages 16 to 98, during three years from 2010 to 2013, with most drivers participating for one to two years. The study sites were located in six states: Florida, Indiana, New York, North Carolina, Pennsylvania, and Washington (WA). Indiana and Pennsylvania were predominantly rural sites, covering about ten counties each; the other four were urban or mixed sites, covering one to three counties. Blanco et al. (2016) [3.1] classified the SHRP2 crash events according to the SHRP2 severity definitions in Appendix G, briefly summarized as follows. Severity Level 1 is a Severe crash with high $|g|$-acceleration, airbag deployment, a likely injury, or some combination of these. Severity Level 2 is a Major PDO crash, and Severity Level 3 is a Minor PDO crash. Severity Level 4 crashes (e.g., a tire bumping a curb) were excluded from the Blanco et al. (2016) [3.1] study because these events are such low severity. (Note that these crash Severity Levels 1–3 should not be confused with SAE Levels 1–5 [3.3] for ADS-equipped vehicles.) Blanco et al. (2016) [3.1] adjusted the SHRP2 crash rates for age. The SHRP2 age categories then had equal proportions to the national driver average age categories. (The original SHRP2 data had been more heavily weighted towards younger drivers, upwardly biasing the original SHRP2 crash rates estimates.)

3.1.3. Google AV versus SHRP2 CV Crash Rates

Blanco et al. (2016) [3.1] compared the Google AV crash data described in Section 3.1.1 to the SHRP2 CV data from 2010 to 2013 [3.1] described in Section 3.1.2. In particular, the VTTI investigators compared the Google AV total crash rate estimate to the SHRP2 total crash rate estimate at three crash Severity Levels. Figure 3.1 is a reproduction of Figure 1 of Blanco et al. (2016) [3.1], showing the main results of that study.

Figure 3.1 plots the total crash rate data by crash Severity Level from Blanco et al. (2016, Figure 1, Table 47) [3.1]. The blue bars are the age-adjusted SHRP2 crash rate estimates. They indicate the human driver age-adjusted CV crash rates in an early version of the SHRP2 database. The green bars indicate the Google AV crash rate estimates in autonomous mode (including transition) in Mountain View, CA, and vicinity from 2009 to 2015. The error bars are the 95% confidence limits estimated by Blanco et al. (2016, Table 47) [3.1].

Blanco et al. (2016, p. iii) [3.1] concluded, based on their data reproduced in Figure 3.1, that "the observed crash rates in the SHRP2 study, at all levels of severity, were higher than the Self-Driving Car rates." For Severity Level 3 crashes, Blanco et al. (2016, p. 41) [3.1] state, "The data also suggest that less-severe events (Severity Level 3 crashes) may happen at a significantly lower rate for self-driving cars."

FIGURE 3.1 **Comparison of CV and AV total crash rate by crash Severity Level. Y-Axis:** Total crash rate (total crashes ÷ total MVMT). **X-Axis:** Crash Severity Level. **Blue bars:** Age-adjusted CV total crash rates of human drivers in the SHRP2 study at six U.S. sites. **Green bars:** Google AV total crash rates in Mountain View and vicinity. Redrawn from Blanco et al. (2016, Figure 1) [3.1]. Data and error bars are from Table 47 in Blanco et al. (2016) [3.1]. The AV and CV fleets were fully instrumented and counted all crashes.

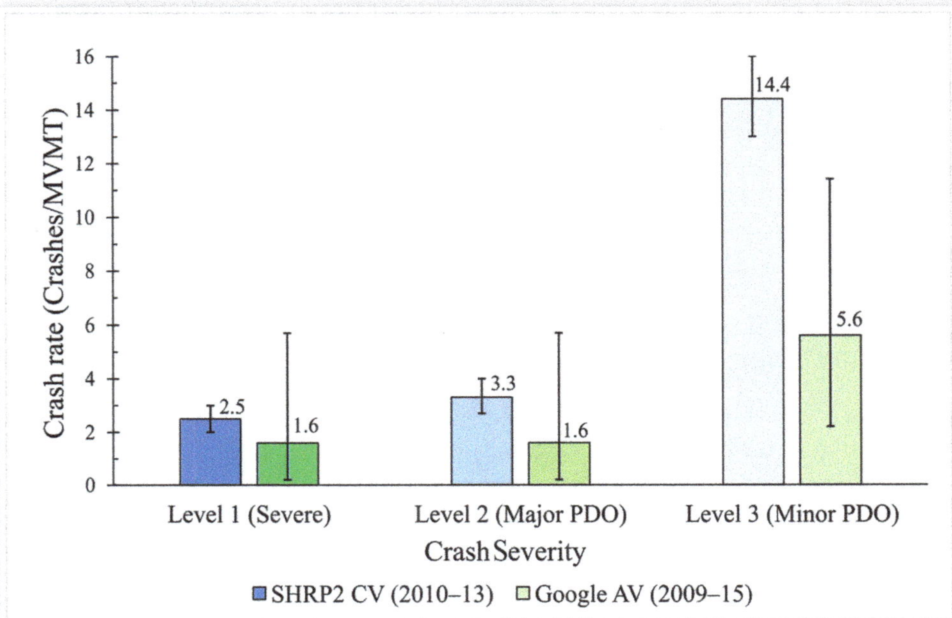

3.2. Epidemiologic Analysis: Blanco et al. (2016) Google AV versus SHRP2

3.2.1. Blanco et al. (2016) Replicated: Google AVs Reduce Crash Rate

To facilitate the critical analysis of Blanco et al.'s (2016) [3.1] original data reproduced in Figure 3.1, Table 3.1 places that original data into the Table 1.1 epidemiologic design for IRR estimation.

Table 3.1 compares the Google AV and age-adjusted SHRP2 CV crash and mileage data from 2009 to 2015, as tabulated by Blanco et al. (2016, Table 47) [3.1], stratified by crash Severity Level. Because the CV crash counts had non-integer values after the age adjustment by Blanco et al. (2016) [3.1], StatsDirect [3.4] was used to estimate each Google AV IRR and its exact mid-P 95% confidence limits and *P*-values.

Table 3.1A estimates a Google AV Severity Level 1 (Severe) crash IRR of 0.64, indicating that prototype Google AVs decreased the Severe crash rate by 36% compared to age-adjusted

TABLE 3.1 Blanco et al. (2016) original data: Epidemiologic analysis.

	Google AV	SHRP2 CV[e]	Google AV[a]	SHRP2 CV[e]
	A. Severe		**B. Major PDO**	
All crashes	2[b]	**85.71**[f]	2[b]	**114.98**[f]
MVMT	1.266611[c]	**34.843**[g]	1.266611[c]	**34.843**[g]
Crash rate	1.6[d]	2.46[h]	1.6[d]	3.30[h]
IRR (95% CI)[i]	0.64 (0.11–2.2), *P* = 0.58		0.48 (0.08–1.6), *P* = 0.30	
	C. Minor PDO		**D. Total crashes**	
All crashes	7[b]	**501.74**[f]	11[b]	**702.43**[f]
MVMT	1.266611[c]	**34.843**[g]	1.266611[c]	**34.843**[g]
Crash rate	5.6[d]	14.40[h]	8.8[d]	20.16[h]
IRR (95% CI)[i]	0.38 (0.17–0.76), *P* = 0.004		0.43 (0.23–0.8), *P* = 0.002	

Google AV crash rates versus age-adjusted SHRP2 CV crash rates, stratified by crash severity.

A **bold** number in a 2×2 subtable above indicates no identified error or bias.

The underlined words in the notes below are revised in later tables.

Notes:

[a] Google AV data from 2009 to October 31, 2015, not standardized.

[b] Google AV crashes in autonomous mode or during a transition (Blanco et al., 2016, Table 4), [3.1], not standardized.

[c] Total Google AV mileage while in autonomous mode as of October 31, 2015 (Blanco et al., 2016, p. 8), not standardized. Mountain View and vicinity, including 56,731 miles in Austin, TX, from July 1, 2015, to October 31, 2015.

[d] Google AV crash rate estimate (crashes per MVMT).

[e] Age-adjusted SHRP2 CV data.

[f] SHRP2 CV crashes adjusted for age using the Blanco et al. (2016, p. 66) [3.1] method.

[g] SHRP2 CV mileage adjusted for age using the Blanco et al. (2016, p. 66) [3.1] method.

[h] SHRP2 CV age-adjusted crash rate estimate (age-adjusted crashes per age-adjusted MVMT) (Blanco et al., 2016, Table 47) [3.1].

[i] The IRR, exact mid-P confidence levels, and exact two-tailed *P* must be estimated with the StatsDirect rate ratio program [3.4] because of the non-integer CV cases (see Chapter 1, Section 1.2).

SHRP2 CVs. Nonetheless, these data, if valid, are also reasonably compatible with this AV rate ranging from an 89% decrease to a 120% increase versus the SHRP2 CV rate.

Table 3.1B estimates a Google AV Severity Level 2 (Major PDO) crash IRR of 0.48, indicating that Google AVs decreased the Major PDO crash rate by 52% compared to age-adjusted SHRP2 CVs. Nonetheless, these data, if valid, are also reasonably compatible with this AV rate ranging from a 92% decrease to a 63% increase versus the SHRP2 CV rate.

Table 3.1C estimates a Google AV Severity Level 3 (Minor PDO) crash IRR of 0.38, indicating that Google AVs decreased the Minor PDO crash rate by 62% compared to age-adjusted SHRP2 CVs. Nonetheless, these data, if valid, are also reasonably compatible with this AV rate ranging from an 83% decrease to a 24% decrease versus the SHRP2 CV rate.

The crash Severity Levels in Tables 2.5A-C are homogeneous ($P = 0.77$), and hence the AV and CV crashes can be separately combined across crash Severity Levels, as shown in Table 3.1D.

Table 3.1D estimates a Google AV total crash IRR of 0.43, indicating that Google AVs decreased the total crash rate by 57% compared to the age-adjusted SHRP2 CVs. Nonetheless, these data, if valid, are also reasonably compatible with the total Google AV crash rate ranging from a 77% to 25% decrease versus the SHRP2 CV total crash rate.

Figure 3.2 plots the Google AV crash IRR estimates and their 95% CIs from Table 3.1, based on the original data of Blanco et al. (2016) [3.1]).

The IRR estimates in Table 3.1 and Figure 3.2 are consistent with, in the absence of other bias or errors, Blanco et al.'s (2016, p. iii) [3.1] main conclusion that "the observed crash

FIGURE 3.2 Epidemiologic analysis of AV and CV crash rates in Blanco et al. (2016) [3.1] stratified by crash Severity Levels 1 to 3. Y-Axis: Google AV Crash IRR estimate. (Google AV crash rate estimate ÷ SHRP2 CV age-adjusted crash rate estimate). **X-Axis**: Crash Severity Level. **Green circles:** IRR estimates for Google AV fleet. **Blue horizontal line:** Age-adjusted SHRP2 CV crash rate comparison (IRR = 1). Crash rate estimates are based on Blanco et al.'s (2016) [3.1] original data, reproduced in this Chapter in Table 3.1 and Figure 3.1.

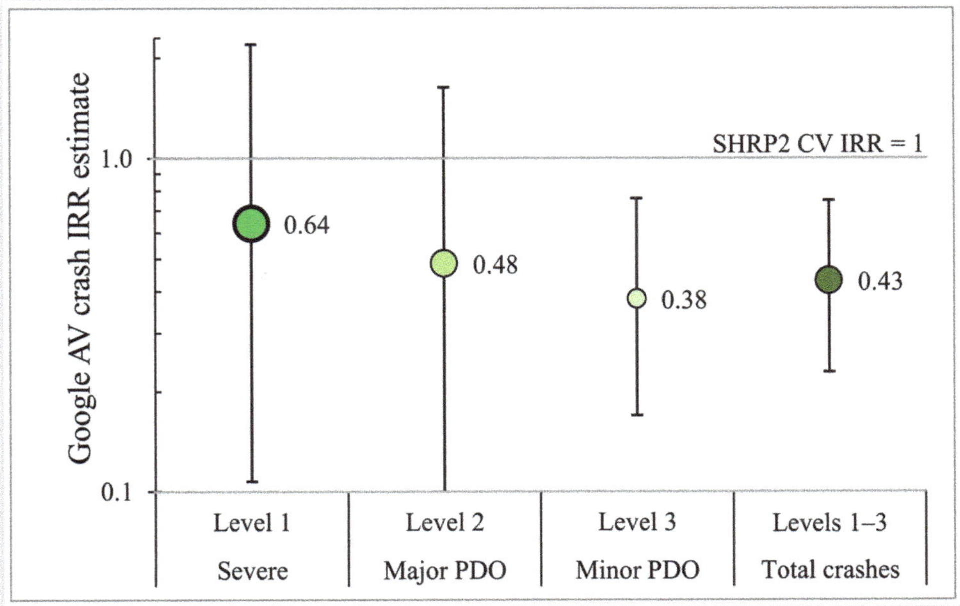

rates in the SHRP2 NDS, at all levels of severity, were higher than the Self-Driving Car rates." In other words, Google AVs have lower crash rates than CVs at all crash Severity Levels.

3.2.2. Blanco et al. (2016) versus Schoettle and Sivak (2015)

Blanco et al.'s (2016) [3.1] conclusion that AVs have a lower crash rate than CVs is precisely the opposite of the conclusion reached by Schoettle and Sivak (2015) [3.2] with the identical 11 Google AV crashes. A different selection of the comparison CV benchmark is the likely reason for the discrepancy: Blanco et al. (2016) [3.1] selected CV crashes from the SHRP2 study, while Schoettle and Sivak (2015) [3.2] selected the 2013 U.S. national police-recorded CV crashes and added specific percentages of injury and PDO unrecorded crashes. The specific unrecorded crash percentages selected by Schoettle and Sivak (2015) [3.2] downwardly biased their 2013 U.S. CV total (recorded + unrecorded) crash rate estimate compared to Blanco et al.'s (2016) [3.1] SHRP2 CV total crash rate estimate.

However, unrecorded crash selection bias could not have biased Blanco et al.'s (2016) [3.1] SHRP2 data analysis described in Section 3.1.3. The reason is that the VTTI investigators did not have to make any assumptions about unknown or ambiguous percentages of unrecorded injury and PDO crashes. Also, *all* CV crashes (whether police recorded or not) were tabulated in the SHRP2 database, just as they presumably were for the Google AV crashes. In other words, assumptions about the percentage of unrecorded crashes could not bias Blanco et al.'s (2016) [3.1] comparison of the Google AV crash rates to the SHRP2 CV crash rates. But unrecorded CV crashes could still have biased Schoettle and Sivak's (2015) [3.2] 2013 U.S. national CV crash rate estimate (see Section 2.2.1).

However, Blanco et al.'s (2016) [3.1] data in Table 3.1, on which Figure 3.2 was based, did contain a site selection bias. Therefore, the conclusions of Blanco et al. (2016) [3.1] about their comparisons of AV to CV crash rates should not be considered valid until the site selection bias is adjusted.

3.3. Critical Analysis of Blanco et al. (2016): Site Selection Bias

A site selection bias is present in Blanco et al.'s (2016) [3.1] data reproduced in Figure 3.1 and Table 3.1. Therefore, the bias is also present in the Google AV IRR estimates in Figure 3.2 based on that data. Recall from Section 2.1.3 that Schoettle and Sivak (2015) [3.2] had correctly warned about site selection bias in comparing AV crash rate estimates to CV crash rate estimates under different environmental conditions. Furthermore, there can also be differences in crash rates between sites for an innumerable list of other reasons. Hence, it is essential to find controls that can adjust for site differences that can differentially affect AV and CV crash rates, confounding any AV IRR estimate.

For example, a site selection bias might still be present in Blanco et al.'s (2016) [3.1] analysis caused by differences between the CV crash rates at the Mountain View site and the crash rates at the SHRP2 CV geographic sites. After removing the 56,731 Google AV miles driven in Austin, TX, from July 1, 2015, to October 31, 2015, all the remaining Google AV crashes and mileage were in Mountain View, CA, and vicinity. As mentioned in Section 3.1.1, the SHRP2 data were collected in six non-CA states, with many rural or mixed urban/rural sites. The SHRP2 data also included highway and city-street driving, whereas the

Google AV data were mostly city-street driving in Mountain View and vicinity. In the SHRP2 CV dataset, highway driving has a substantially lower crash rate than urban driving, at all crash Severity Levels (Blanco et al., 2016, pp. 27–28, Figures 7–9) [3.1].

3.3.1. Site Selection Bias Does Not Explain Why Google AVs Reduce the Severe Crash Rate (Table 3.1A)

Is the Google AV Severe crash rate reduction effect versus SHRP2 CVs in Figure 3.1 (leftmost bars) a valid AV effect, as concluded by Blanco et al. (2016) [3.1]? This conclusion appeared to be confirmed by the Google AV Severe crash IRR estimate of 0.64 in Table 3.1A, a 36% reduction in the Severe crash rate by AVs compared to SHRP2 CVs. The conclusion is also consistent with the leftmost data point in Figure 3.2, indicating a Google AV IRR estimate of 0.64. However, the Google AV reduction in the Severe crash rate can only be declared a valid effect if all biases are removed. For example, a site selection bias could create a false appearance that Google AVs reduced Severe crashes when they didn't. Figure 3.3 helps answer whether this crash rate reduction effect was confounded by site selection bias.

Figure 3.3A replots the Blanco et al. (2016) [3.1] Severe crash rate data from Figure 3.1 (leftmost bars) and Table 3.1A. Figure 3.3A illustrates the observation made in Section 3.2.1 that the Google AV fleet Severe crash rate estimate of 1.6 (green bar) was a 36.0% reduction compared to the age-adjusted SHRP2 CV fleet Severe crash rate estimate of 2.5 (blue bar).

FIGURE 3.3 The Severe crash rate reduction by the Google AV fleet was not likely caused by site selection bias in the Blanco et al. (2016) [3.1] analysis. A. Y-Axis: Severe crash rate estimate (total Severe crash estimate ÷ total MVMT estimate). The Google AV fleet (green bar) reduced the Severe crash rate estimate in Mountain View and vicinity by 36% compared to the SHRP2 CV sites (blue bar). (Redrawn from Figure 3.1, leftmost bars.) **B. Y-Axis:** The police-recorded CV injury crash rate. The injury crash rate in Mountain View (right blue-hatched bar, 0.68) was 14.7% higher than at the six SHRP2 sites (left blue-hatched bar, 0.58) in the same period.

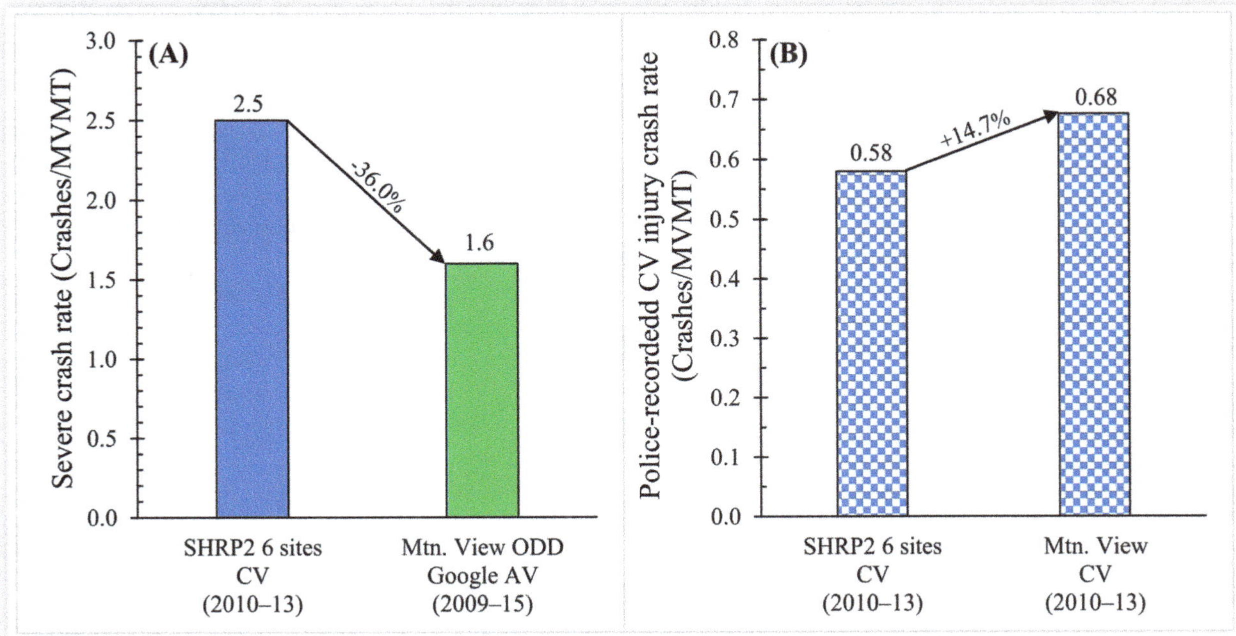

The Google AV and SHRP2 naturalistic driving estimates in Figure 3.3A include all police-recorded and unrecorded Severe crashes. Therefore, unrecorded crash selection bias from overestimates of unrecorded CV crashes cannot explain the higher CV versus AV rate in Figure 3.3A.

However, can site selection bias explain the 36% reduction in the Google AV Severe crash rate in Figure 3.3A? It could if Mountain View (the principal operating design domain or ODD for the Google AV fleet) had about a 36% lower CV injury crash rate than at the SHRP2 sites. If so, then Google AVs would not have caused the 36% AV reduction in Severe crash rate versus the SHRP2 CVs in Figure 3.3A. Instead, the 36% Severe crash reduction effect could be entirely caused by site selection bias. There might simply be a generally lower risk of a Severe crash in Mountain View than at the SHRP2 sites (for example, milder weather in Mountain View compared to the SHRP2 sites).

Figure 3.3B investigates whether such site selection bias can explain the AV Severe crash rate reduction effect in Figure 3.3A. Figure 3.3B (right blue-hatched bar) is the police-recorded injury crash rate for CVs in Mountain View from 2010 to 2013. The numerator of that injury crash rate is the count of CV injury crashes on Mountain View city streets in the SWITRS database [3.5] from 2010 to 2013. (Note that the query for the SWITRS database [3.5] for Mountain View was chosen to include crashes only on city streets under Mountain View police department jurisdiction. Thus, the search excluded crashes on state highways, freeways, interstates, or other roads under CHP jurisdiction in Mountain View city limits.) The denominator of that injury crash rate was the CV mileage on Mountain View city streets in the HPMS database [3.6]. (Note that this mileage estimate does not have to be selected as "city streets only" for Mountain View. The HPMS database [3.6] already excludes mileage on highways and freeways under CHP jurisdiction within a city. The exception is for a city that is also a county, such as San Francisco.) In short, the method used here for estimating the Mountain View injury crash rate ensured that the numerator and denominator of the crash rate had matched driving locations.

Figure 3.3B (left blue-hatched bar) is the police-recorded CV injury crash rate at the six SHRP2 sites during regular non-instrumented vehicle driving for the same years 2010-2013 (based on data from Blanco et al., 2016 [3.1], their Tables 24, 27, 30, 33, 38, and 41). The numerator is all police-recorded crashes on city streets and highways at the six SHRP2 sites. The denominator is all mileage on city streets and highways at the six SHRP2 sites. Therefore, the numerator and denominator had matched driving locations. (Note that the CV injury crash data in Figure 3.3B are from state and county police-recorded crash databases and are independent of the Google AV and the SHRP2 CV crash rate data in Figure 3.3A.) In sum, Figure 3.3B shows that the estimate for the police-recorded CV injury crash rate is 14.7% higher in Mountain View, CA (right blue-hatched bar, 0.68) than at the six SHRP2 sites (left blue-hatched bar, 0.58).

That 14.7% increase in the CV injury crash rate estimate in Mountain View compared to the SHRP2 sites indicates that the 36% Google AV Severe crash rate decrease in the Mountain View area (Figure 3.3A) could not have been caused by site selection bias. The Google AV Severe crash rate reduction is a valid effect (in the absence of other bias). In other words, the reduction in the Severe crash rate by the Google AV fleet cannot be dismissed as being caused by site selection bias.

In other words, if the data in Figure 3.3B had shown a lower CV injury crash rate in Mountain View than at the SHRP2 sites, then that lower rate could have confounded the 36% Google AV Severe crash rate reduction seen in Figure 3.3A. However, Figure 3.3B shows the opposite: Mountain View had a 14.7% higher CV injury crash rate than the SHRP2 sites during much of the Google AV test period. It follows that the 36% Google AV Severe crash rate reduction effect in Figure 3.3A cannot be caused by a site selection bias (from CV crash rate differences between the Mountain View site and the SHRP2 sites).

In conclusion, Google AVs cause a valid reduction of Severe crash rates versus SHRP2 CVs (Figure 3.3A) that is not caused by site selection bias (Figure 3.3B).

Figures 3.3A and 3.3B, considered jointly, predict that an unbiased Google AV Severe crash reduction effect should even be greater than the 36.0% decrease seen in Figure 3.3A. The reason is that the 36.0% decrease (Figure 3.3A) occurred in the presence of a 14.7% increase in the general probability of an injury crash in Mountain View during much of the same period (Figure 3.3B). This prediction of a better protective effect of Google AVs for Severe crashes than is apparent in Figure 3.3A is quantitatively tested by adjusting the AV crash rates using a standardization method in Section 3.4.

3.3.2. Site Selection Bias Explains Why Google AVs Appear to Reduce the Major PDO Crash Rate (Table 3.1B)

Figure 3.4 helps answer whether site selection bias reduced the Google AV Major PDO crash rate compared to the SHRP2 Major PDO crash rate evident in Figure 3.1 (middle bars) and Table 3.1B. Figure 3.2 shows the Google AV Major PDO IRR estimate was 0.48, a protective effect.

FIGURE 3.4 The Major PDO crash rate reduction by the Google AV fleet was likely caused by site selection bias in the Blanco et al. (2016) [3.1] analysis. A. Y-Axis: Major PDO crash rate estimate (total Major PDO crash estimate ÷ total MVMT estimate). The Google AV fleet (green bar) reduced the Major PDO crash rate estimate in Mountain View and vicinity by 51.5% compared to the SHRP2 CV sites (blue bar). (Redrawn from Figure 3.1, middle bars.) **B. Y-Axis:** The police-recorded CV Major PDO crash rate. The police-recorded Major PDO crash rate in Mountain View (right blue-hatched bar, 0.57) was 52.9% lower than at the six SHRP2 sites (left blue-hatched bar, 1.21) in the same period.

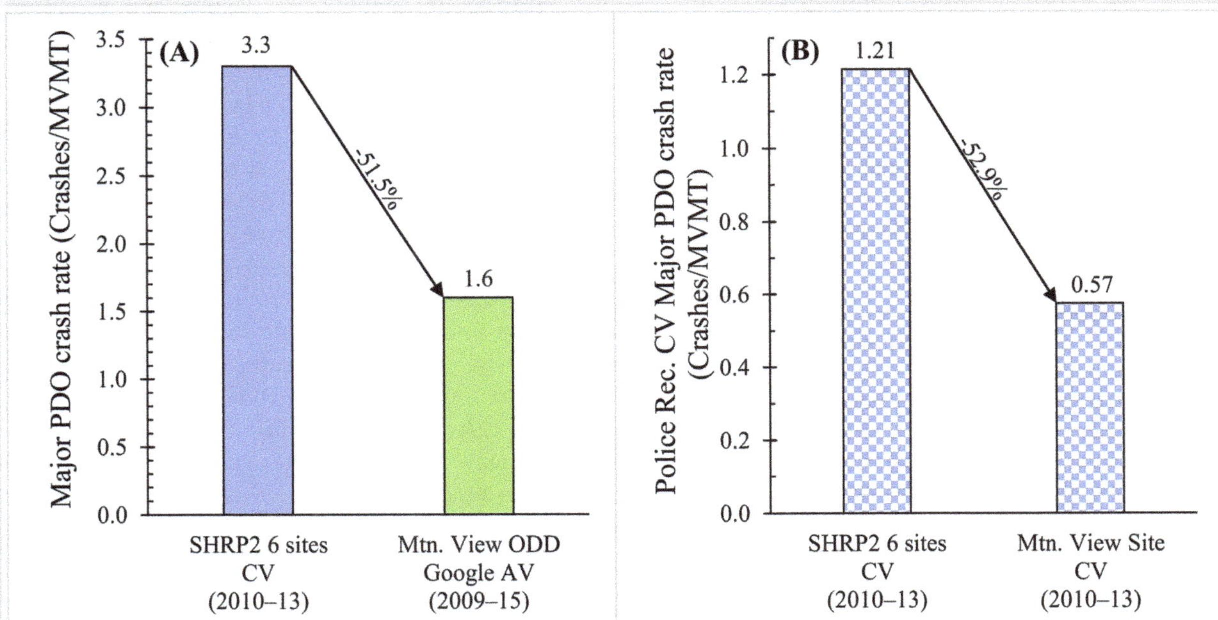

Figure 3.4A replots the crash rate data in Figure 3.1 (middle bars) from Blanco et al. (2016) [3.1] for Severity Level 2 crashes (Major PDO). Blanco et al.'s (2016) [3.1] data in Figure 3.4A shows that the Major PDO crash rate estimate for the Google AV fleet (green bar, 1.6 crashes per MVMT) was 51.5% lower than that for the SHRP2 CV fleet during naturalistic driving (blue bar, 3.3 crashes per MVMT). (Note that Figure 3.4A represents all Major PDO crashes, both police-recorded and not police-recorded, for both the Google AV fleet and the SHRP2 CV fleet, so again there is no unrecorded crash selection bias.)

However, the Major PDO AV crash rate reduction effect of 51.5% in Figure 3.4A could still arise from site selection bias if Mountain View had a lower Major PDO CV crash rate estimate than the Major PDO CV crash rate estimate for the SHRP2 sites. In other words, whether vehicles were exposed to autonomous mode may have nothing to do with the seeming 51.5% reduction in Major PDO crashes for AVs versus CVs in Figure 3.4A. Instead, the reduction may arise entirely from a lower Major PDO CV crash rate in Mountain View versus the SHRP2 sites.

Figure 3.4B tests this site selection bias issue for Major PDO crashes. The right blue-hatched bar is the police-recorded Major PDO CV crash rate estimate in Mountain View, CA, for the combined years of 2010-2013. The left blue-hatched bar is the police-recorded Major PDO CV crash rate estimate for the same combined years of 2010–2013 at the six SHRP2 sites (data from Blanco et al., 2016, Tables 24, 27, 30, 33, 38, and 41) [3.1]. (Note that the data plotted in Figure 3.4B are from state and county police-recorded crash databases, which are independent of the crash data used for the Google AV and SHRP2 naturalistic driving CV crash rates in Figure 6A.) Figure 3.4B shows by visual inspection that the police-recorded Major PDO CV crash rate is 52.9% lower in Mountain View, CA (right bar, 0.57) than at the SHRP2 sites (left bar, 1.21).

This 52.9% lower estimate of the Mountain View police-recorded CV Major PDO crash rate in Figure 3.4B indicates that the appearance of a 51.5% AV reduction in the Major PDO crash rate in Figure 3.4A is not a valid AV effect. Instead, the reduction is caused by site selection bias. In other words, the appearance of an AV Major PDO crash rate decline in Figure 3.4A can be wholly attributed to the fact that Mountain View had a 52.9% lower Major PDO CV crash rate than did the SHRP2 sites (Figure 3.4B) during most of the Google AV test period. On the contrary, a higher or equal Major PDO CV crash rate for Mountain View versus the SHRP2 sites would be required to prove that site selection bias did not cause the seeming reduction in Major PDO crashes by Google AVs. In other words, it cannot be denied that the Google AV Major PDO crash rate in Figure 3.4A (green bar) is 51% lower than the SHRP2 Major PDO crash rate (blue bar). However, this Major PDO crash rate reduction is not attributable to Google AVs; it is wholly an artifact of site selection bias.

To restate for clarity, the site selection bias shown in Figure 3.4B is that Mountain View exhibits a 52.9% lower police-recorded Major PDO CV crash rate (right bar, 0.57) than the SHRP2 sites (left bar, 1.21). Hence, the apparent 51.5% Major PDO crash rate reduction by the Google AV fleet in Figure 3.4A can be wholly attributed to the 52.9% Major PDO CV crash rate reduction in Figure 3.4B for the Mountain View site compared to the six SHRP2 sites. In other words, the apparent decrease in the AV Major PDO crash rate from 3.3 to 1.6 in Figure 3.4A is not a valid Google AV crash rate reduction effect. Instead, the reduction is more succinctly explained because Mountain View has a substantially lower Major PDO CV crash rate than the CVs at the SHRP2 sites (Figure 3.4B).

Note that the absolute sizes of the Major PDO crash rates in Figures 3.4A and 3.4B are considerably different because only police-recorded Major PDO crashes were counted in Figure 3.4B. But all Major PDO crashes were counted in Figure 3.4A. However, the different absolute sizes of the crash rates in the two panels A and B in Figure 3.4 are irrelevant to the similarity in the direction and degree of the percentage decline in the two panels.

In conclusion, Google AVs do not cause a valid 51.5% reduction of Major PDO crash rates versus SHRP2 CVs (Figure 3.4A). Site selection bias was wholly responsible for the Major PDO Google AV crash rate reduction effect in Figure 3.4A. This conclusion will be quantitatively confirmed after adjusting the Google AV Major PDO crash rate using a standardization method in Section 3.4.

3.3.3. Site Selection Bias Explains Why Google AVs Appear to Reduce the Minor PDO Crash Rate (Table 3.1C)

Minor PDO crashes often do not have police records entered in state databases by many police jurisdictions, so one cannot use police-recorded data to test for and remove possible site selection bias. However, Mountain View has a substantially lower Major PDO police-recorded CV crash rate than the SHRP2 sites (Figure 3.4B). Hence it is plausible that Mountain View may also have a considerably lower Minor PDO CV crash rate than the SHRP2 sites. Note the seeming substantial reduction of the Minor PDO crash rate by AV exposure (Figure 3.1, rightmost green bar, 5.6) compared to the Severity Level 3 Minor PDO SHRP2 CV crash rate (Figure 3.1, rightmost blue bar, 14.4). However, this reduction was almost certainly caused by site selection bias, as was the Major PDO crash rate reduction (see Section 3.3.2). Therefore, it is concluded that the apparent substantial decline in Minor PDO crashes in Mountain View and vicinity from AV exposure compared to the SHRP2 CVs (Figure 3.1, rightmost bars) may reflect nothing more than that Mountain View had a lower Minor PDO CV crash rate than the six SHRP2 sites, as was the case for the Major PDO AV crashes (Section 3.3.2). This conclusion will be quantitatively confirmed after adjusting the AV Minor PDO crash rate using a standardization method in Section 3.4.

3.4. Standardizing the AV Crash Data Adjusts for Site Selection Bias

Standardization is defined as "A set of techniques, based on weighted averaging, used to remove as much as possible the effects of differences in age or other confounding variables in comparing two or more populations" (Porta, 2014, p. 268) [3.7]. Standardization is a well-established epidemiologic method to reduce the effect of site selection bias, such as when comparing disease mortality rates between different countries with different population ages. The data from each country are standardized to a common population to control for age differences between countries. For example, assume country A has a much older population than country B on average. Then country A (with the disease) could (falsely) appear to have a higher mortality rate than country B (without the disease) simply because country A has an older population, not because it has the disease. For further explanation, see any introductory epidemiology textbook such as Rothman (2012, pp. 188–192) [3.8], or for more advanced treatment, see Rothman et al. (2008, pp. 49–50, 67–69, 265–269, 386–388, 442–446) [3.9].

3.4.1. Standardize Google AV Crashes to SHRP2 Data Sites

Section 3.2.2 investigated qualitatively whether site selection bias could explain the Blanco et al. (2016) [3.1] AV crash rate reduction effects in Figure 3.1. One can also quantitatively adjust

TABLE 3.2 Method to adjust for site selection bias: Standardization of the Google AV crash rate.

	Std. AV[a]	SHRP2 CV[e]	Std. AV[a]	SHRP2 CV[e]
	A. Severe		**B. Major PDO**	
Crashes	**1.71**[b]	**85.71**[f]	**4.25**[b]	**114.98**[f]
MVMT	**1.20988**[c]	**34.843**[g]	**1.20988**[c]	**34.843**[g]
Crash rate	1.41[d]	2.46[h]	3.51[d]	3.30[h]
IRR (95% CI)[i]	0.57 (0.11-2.30), P = 0.64		1.06 (0.31-2.47), P = 0.94	
	C. Minor PDO		**D. Total crashes**	
Crashes	**14.86**[b]	**501.74**[f]	**20.81**[b]	**702.43**[f]
MVMT	**1.20988c**	**34.843**[g]	**1.20988**[c]	**34.843**[g]
Crash rate	12.28[d]	14.40[h]	17.20[d]	20.16[h]
IRR (95% CI)[i]	0.85 (0.50-1.40), P = 0.59		0.85 (0.54-1.30), P = 0.51	

Comparison of the standardized Blanco et al. (2016) [3.1] Google AV crash rates to the age-adjusted SHRP2 CV crash rates, stratified by crash severity.

A **bold** number in a 2×2 subtable means no identified error or bias.

Notes:

[a] Google AV CA data from 2009 to October 31, 2015, standardized by the ratio of Mountain View CV crashes to six SHRP2 sites non-instrumented CV crashes (see Appendix H).

[b] Standardized Google AV CA crashes in autonomous mode or during a transition (Blanco et al., 2016, Table 4) [3.1].

[c] Total Google AV CA mileage while in autonomous mode as of October 31, 2015 (Blanco et al., 2016, p. 8) [3.1]. Mountain View and vicinity only, excluding 56,731 miles in Austin, TX, from July 15, 2015, to October 31, 2015.

[d] Standardized Google AV crash rate estimate (crashes per MVMT).

[e-i] Same as Table 3.1.

for site selection bias using standardization as defined in the paragraph above. Appendix H gives the details of the method used to standardize the Google AV crashes and crash rates.

The Google AV standardized crash rate estimates would be predicted to occur if the AV tests had been conducted at the SHRP2 sites rather than in Mountain View and vicinity. That is, the Google AV test results are standardized to the SHRP2 sites. Figure 3.5 plots, stratified by crash Severity Level, the standardized Google AV rates (green-striped bars) for visual comparison to the same SHRP2 CV rates (solid blue bars) from Figure 3.1.

The standardization method in Appendix H gave rise to the standardized AV rates plotted in Figure 3.5. The standardization of the crude Google AV crash rates removes the site selection bias described in Section 3.2.2. The standardized IRR estimates in Table 3.2 are then obtained after dividing the standardized AV crash rate (cell d) by the age-adjusted SHRP2 CV rate (cell h). Table 3.2 gives the resulting standardized incidence rate ratio (sIRR) estimates (cell i) and confidence intervals, stratified by crash Severity Level.

Table 3.2 allows a quantitative comparison of the standardized Google AV crash rates to the SHRP2 CV crash rates, stratified by crash Severity Level. The standardization method (Appendix H) adjusts for site selection bias because the CV crash rates at the Mountain View site have been standardized to the CV crash rates at the SHRP2 sites as if the Google AV tests had been conducted at the SHRP2 sites.

Severe crashes. Table 3.2A estimates a standardized Google AV Severity Level 1 (Severe) crash IRR of 0.57, indicating that Google AVs decrease the Severe crash rate by 43% versus SHRP2 CVs after adjusting for site selection bias. Nonetheless, these data, if valid, are also reasonably compatible with that rate ranging from an 89% decrease to a 130% increase of the SHRP2 CV Severe crash rate of 0.57. This Google AV standardized Severe crash IRR

FIGURE 3.5 Comparison of CV and standardized AV total crash rates by crash Severity Level. Legend same as Figure 3.1, except the green-striped bars are standardized Google AV total crash rate estimates.

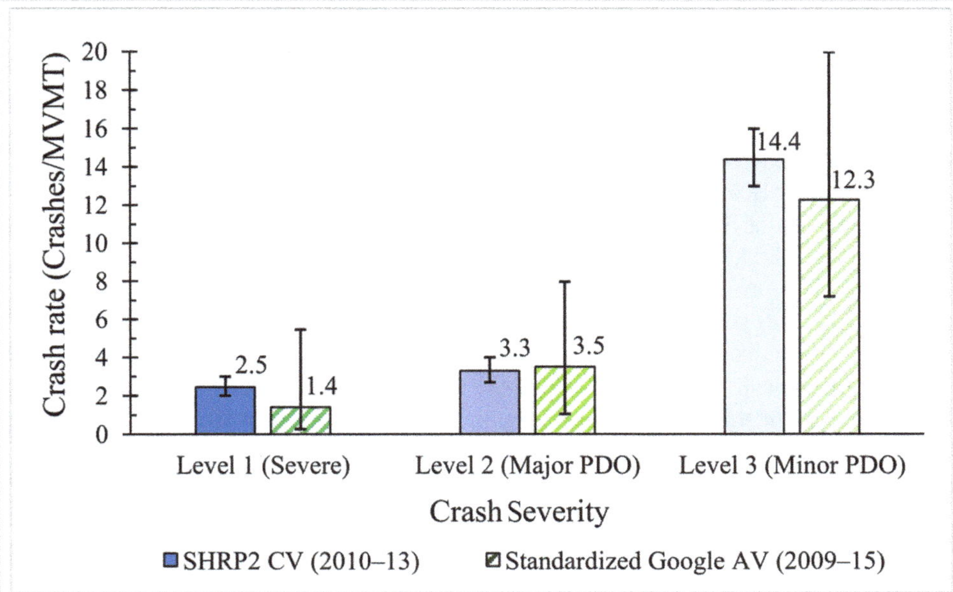

estimate of 0.57 (Table 3.2A) is slightly lower (i.e., more protective) than the Google AV crude Severe crash IRR estimate of 0.64 (Table 3.1A), as predicted at the end of Section 3.3.1.

This 43% quantitative improvement in the Severe crash rate after standardization confirms the visual observation from Figure 3.3 in Section 3.3.1. Namely, the 36% improvement in Severity Level 1 crash rates by Google AV exposure (Figure 3.3A) is a valid effect. It cannot arise from site selection bias because the CV injury crash rate in Mountain View (0.68) is higher than at the SHRP2 sites (0.58) (see Figure 3.3B). In fact, after adjustment for site selection bias, the 43% Google AV reduction in the standardized Severe crash rate is now even more protective than the 36% AV reduction in Severe crashes in the Blanco et al. (2016) [3.1] crude data (Figure 3.3A). In other words, after standardization to adjust for site selection bias, the Google AV test fleet decreased the Severe crash rate by 43% compared to CVs.

This 43% reduction of Severe crashes by the Google AV fleet after standardization is even more protective than the 36% protective effect in the crude Google AV data as predicted in Section 3.3.1 (last paragraph). The prediction was because the 36.0% Severe crash rate decrease (Figure 3.3A) occurred in the presence of a 14.7% increase in the general probability of an injury crash in Mountain View (Figure 3.3B). After adjustment for that general probability increase by standardization, the Google AV reduction in the Severe crash rate became even more pronounced, confirming the prediction.

Major PDO crashes. Table 3.2B estimates a standardized Google AV Severity Level 2 (Major PDO) crash IRR of 1.06, indicating that Google AVs now increase Major PDO crashes by 6% versus SHRP2 CVs. Nonetheless, these data, if valid, are also reasonably compatible with the Google AV Major PDO crash rate ranging from a 69% decrease to a 147% increase of the SHRP2 CV Major PDO crash rate of 3.30.

This increase and reversal of the direction (from protective to causative) of the Google AV standardized Major PDO crash IRR to 1.06 (Table 3.2B) compared to the Google AV crude Major PDO crash IRR estimate of 0.48 (Table 3.1B) confirms the earlier result illustrated in Figure 3.4. Namely, the Google AV reduction of Major PDO crashes (Figure 3.4A) is not a valid AV effect. Instead, the seeming reduction is caused entirely by site selection bias from the 52.9% lower Major PDO CV crash rate in Mountain View (0.57) versus the SHRP2 sites (1.21) (Figure 3.4B).

Minor PDO crashes. Table 3.2C estimates a standardized Google AV Severity Level 3 (Minor PDO) crash IRR of 0.85, indicating that Google AVs decrease Minor PDO crashes by 15% versus SHRP2 CVs. Nonetheless, these data, if valid, are also reasonably compatible with this standardized Minor PDO AV crash rate ranging from a 52% decrease to a 35% increase versus SHRP2 CVs.

The increase of the standardized Minor PDO IRR estimate to 0.85 from the crude Minor PDO crash IRR estimate of 0.38 (Table 3.1) suggests that the Google AV reduction of the Minor PDO crash rate in Figure 3.1 from 14.4 to 5.6 is not a valid AV crash reduction effect. Instead, it is similar to the false Google AV crash reduction effect for Major PDO crashes. That is, the seeming PDO crash rate decline from Google AVs nearly vanishes after stratification. Indeed, stratification raised the Google AV Minor PDO crash rate from 5.6 (Figure 3.1, rightmost green bar) to 12.3 (Figure 3.5, rightmost green-striped bar). Thus the seeming Google AV reduction of the Minor PDO crash rate in Figure 3.1 (rightmost bars) is an artifact of site selection bias, similar to the Google AV Major PDO crash rate reduction.

Total Crashes. The Google AV standardized IRR estimates depicted in Tables 3.2A-C are homogeneous across Severity Levels (*P*-value for homogeneity equals 0.75). Therefore, the AV and CV crashes could be separately summed across crash Severity Levels to form Table 3.2D.

Table 3.2D estimates a standardized total Google AV IRR of 0.85, indicating that Google AVs would be expected to decrease overall crashes by 15% versus SHRP2 CVs if the Google AV tests had been conducted at the SHRP2 sites instead of in Mountain View, CA, and vicinity. Nonetheless, these data, if valid, are also reasonably compatible with the overall Google AV crash rate ranging from a 44% decrease to a 26% increase versus SHRP2 CVs.

Figure 3.6 summarizes the Google AV standardized IRRs (squares) from Table 3.2, stratified by crash severity. In addition, Figure 3.6 plots the Google AV crude IRRs (green circles) from Figure 3.2 for comparison.

Severe Crashes. The plot in Figure 3.6 (next page) shows that the two leftmost large symbols representing Severe crash indicate that standardizing the Google AV Severe crash rate reduces the original AV Severe crash IRR from 0.64 (large green circle) to the standardized Severe crash IRR estimate of 0.57 (large square with horizontal stripes). This result confirms the visual impression from Figure 3.3 that Google AV exposure has a valid effect in reducing Severe crash rates relative to CVs. Figure 3.6 (Severe crash) shows that the AV Severe crash reduction effect in Figure 3.3A remains present after removing site selection bias by standardization (large square). After removing the site selection bias, the AV Severe crash reduction effect becomes even more protective, decreasing from 0.64 to 0.57. This quantitative result confirms the qualitative result in Section 3.3.1 that the Google AV Severe crash rate reduction is a valid AV exposure effect. The final Severe crash standardized IRR of 0.57 (after adjustment for site bias) indicates that the Google prototype AV fleet reduced the Severe crash rate by 43% compared to human-driven CVs.

Major PDO Crashes. Contrary to Severe crashes, Figure 3.6 shows that the Major PDO standardized crash IRR estimate (medium size square) reversed direction. The crude AV Major PDO crash IRR estimate (medium size green circle) was 0.48. After standardization, the adjusted AV Major PDO crash IRR estimate was 1.06. In other words, after removing

FIGURE 3.6 **Crude versus standardized Google AV IRR estimates stratified by crash Severity Level. Y-Axis:** Google AV Crash IRR. **X-Axis:** Crude versus standardized IRR estimates by crash Severity Level. **Filled green circles:** Crude Google AV IRR estimates (Figure 3.2 and Table 3.1, note i) based on original data of Blanco et al. (2016) [3.1]. **Squares:** Standardized AV IRR estimates (Table 3.2, note i). **Blue horizontal line:** SHRP2 CV comparison, crash IRR = 1.

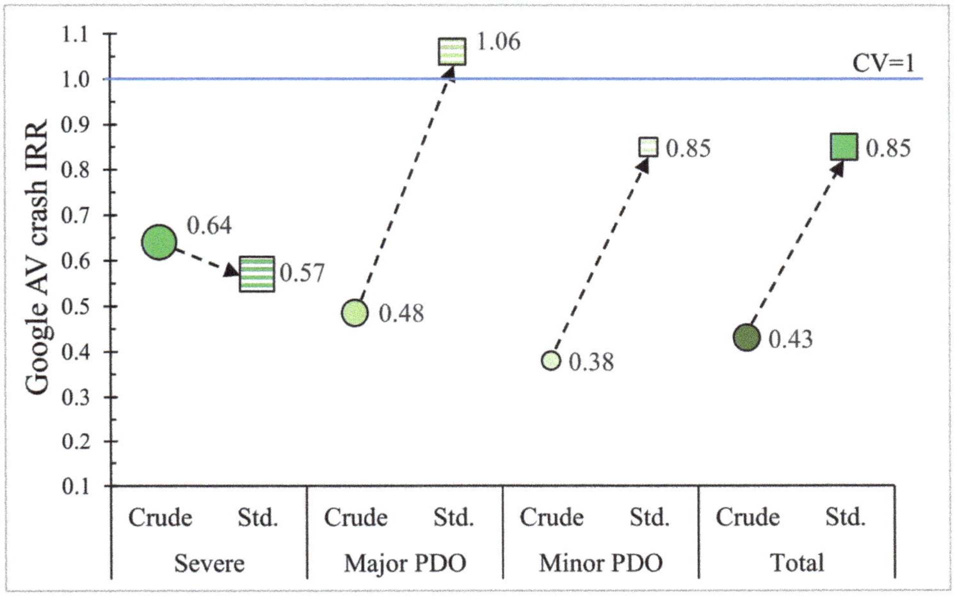

the site selection bias by standardizing the Google AV PDO crash rates in Mountain View to the SHRP2 sites, there is no longer a Google AV Major PDO crash reduction effect. Thus, standardization numerically proves what was suggested by visual inspection of Figure 3.4. the original Blanco et al. (2016) [3.1] Major PDO crash reduction effect (crude IRR = 0.48, Figure 3.2, Table 3.1B) was an artifact of site selection bias.

Minor PDO Crashes. The Minor PDO standardized crash IRR (squares) moved closer to IRR = 1, or the SHRP2 CV crash rate. Standardization has removed the original Blanco et al. (2016) [3.1] Minor PDO crash reduction effect in the Blanco et al. (2016) [3.1] crude crash rate data. This removal was shown in Figure 3.1 (right side) and the Minor PDO crude IRR in Figure 3.2 (crude IRR = 0.38, small green circle). That is, Google AV Severity Level 3 (Minor PDO) no longer shows any AV crash reduction effect (adjusted IRR = 0.85, Figure 3.6, small green striped square).

Hence, the Google AV crash reduction effect for Major and Minor PDO crash rates visible in Figures 3.1 and 3.2 is more plausibly explained as an artifact caused by site selection bias. The bias arises from the 52.9% lower incidence of PDO CV crashes in the city of Mountain View, CA, versus PDO CV crashes at the six SHRP2 sites (Figure 3.4B). In short, Google AV exposure in Mountain View and vicinity does not reduce PDO crashes after adjustment for site selection bias. Thus the quantitative PDO IRR standardization results in Figure 3.6 (squares) and Table 3.2 confirm the qualitative result from visual inspection of Figure 3.4 as described in Section 3.3.2. Namely, AV exposure has little or no effect on PDO crash rates after removal of site selection bias by standardization.

3.4.2. Summary and Discussion of Standardized Results

CV crash rate differences between the Mountain View and SHRP2 sites were adjusted by standardization (see Section 3.4.1 and Appendix H). After standardization, Table 3.2A and Figure 3.6 (leftmost square) indicate that the Google AV data examined by Blanco et al. (2016) [3.1] had a Severity Level 1 (Severe) standardized crash IRR estimate of 0.57. That IRR indicates that Google AVs would have reduced the Severe crash rate by 43% if the Google test had been conducted at the SHRP2 sites. Major PDO crashes would be increased by 6%, and Minor PDO crashes would be decreased by 15%. The standardized crash severity strata were homogeneous so that they could be validly combined. Google AVs reduce total crashes by about 15% compared to CVs.

Standardization removed site selection bias, giving rise to more valid estimates of the Google AV crash rate at each crash Severity Level. The standardization results indicated that Google AVs reduced the Severe crash rate by 43% but increased the Major PDO crash rate by 6%. That observation suggests the new hypothesis that Google AVs convert a Severe crash into a Major PDO crash by not exceeding posted speed limits as much as human drivers do. In other words, this converting of Severe crashes involving injuries, etc., into Major PDO crashes is similar to what would be expected to happen if human drivers, like AVs, had a zero incidence of exceeding the posted speed limit by more than 10 or 15 miles an hour. (For a discussion of this new speed hypothesis, see Section 9.6.)

3.4.3. Limitations of Standardization

It is important to note that there is a limit to how far standardization can adjust for site differences. The standardization method compares the CV crashes at one site to another standard site and uses that standardization factor to adjust the AV crashes to that expected if the AV test had been conducted at the standard site. However, there could be subtle "interactions" between AV behavior and site factors. The simple standardization method conceptualizes "sites" as a "simple main effect" on behaviors or factors contributing to crashes. It does not account for site-specific factors that might interact with AV behavior (differently from CV behavior). For example, a site-specific factor would occur if a particular company's AV fleet had difficulty with left turns at intersections. The AV test site may have featured many more intersections than the CV standard site. Then the standardization method used here may not be sufficient to handle all site differences with a single standardization. Despite this limitation, it is still essential to perform standardization (or other methods) to adjust for site selection bias arising from CV crash rate differences. The limitation is that just one standardization may not entirely adjust for all site differences.

A second limitation of the standardization method applied here is that it standardizes the Google AV Mountain View data only to the six SHRP2 sites. That does not necessarily generalize to other sites. The SHRP2 sites have a site selection bias when compared, for example, to the U.S. national data. Indeed, crash reporting requirements vary considerably by state (Blanco et al., 2016, pp. 6–8) [3.1]. A separate standardization would be required to standardize the Google AV data to the U.S. national site. However, any standardization to a national, state, county, or city site with only police-recorded crash counts (unlike the SHRP2 sites that counts all crashes) would still be confounded by unrecorded crash selection bias as described in Section 3.5 below.

3.5. Blanco et al. (2016): Google AV versus U.S. National CV Comparison is Confounded by Unrecorded Crashes

The CV unrecorded crash issue when comparing CV to AV crash data was first raised by Schoettle and Sivak (2015) [3.2] (Section 2.1.2). This issue was a central one in the Blanco et al. (2016) [3.1] study, which did a thorough investigation of unrecorded crashes in addition to the comparison to the SHRP2 data (Sections 3.1, 3.2, 3.3, 3.4). Blanco et al.'s (2016) [3.1] data and analysis are reviewed in Section 3.5.1, and the general issue of unreported and unrecorded crashes is discussed further in Appendix E.

3.5.1. Unrecorded Crash Selection Bias: SHRP2 versus U.S. National

Figure 3.1 reproduced the Blanco et al. (2016, Figure 1) [3.1] results comparing the Google AV crash rate estimates to the benchmark SHRP2 CV crash rate estimates. Again, there was no unrecorded crash selection bias because the AV and CV crash rate estimates at every crash Severity Level included all crashes (both police-recorded and not police-recorded).

However, unrecorded crashes become critical when comparing AV (or naturalistic driving) crash rates to police-recorded crash rates in public databases. Unrecorded crashes create a substantial crash selection bias because of the uncertain number of unrecorded crashes, as first noted by Schoettle and Sivak (2015) [3.2] (Section 2.1.2 and Appendix E). Blanco et al. (2016) [3.1] extensively investigated the unrecorded crash issue, as described in the next paragraph.

To explore the unrecorded crash issue, Blanco et al. (2016, Figure 2) [3.1] compared the Google AV crash rate estimates to the 2013 U.S. national average CV crash rate [3.10]. That comparison is identical to the benchmark comparison used by Schoettle and Sivak (2015) [3.2] in their original data presented in Chapter 2, Table 2.2. Thus, Blanco et al. (2016, Figure 2) [3.1] correctly recognized, as did Schoettle and Sivak (2015) [3.2] (Section 2.1.2), that, in comparing the AV crash rate to the national CV crash rate, it is critical to estimate the unrecorded crashes in the national crash data. Otherwise, there would be a substantial upward crash selection bias from comparing <u>all</u> AV crashes (police-recorded and unrecorded) to <u>only</u> police-recorded U.S. national CV crashes. As a result, the AV crash rate would appear absurdly high relative to the CV crash rate for reasons entirely unrelated to AV exposure. That is, the AV crash rate would appear to be falsely high simply because of the differential way crashes were selected for the AV and CV groups.

Blanco et al. (2016) [3.1] correctly recognized that it was nontrivial to estimate the unrecorded crashes for everyday driving without fully-instrumented vehicles that record all crashes. There is a wide variation in unrecorded crash percentages in different police jurisdictions and various studies of unrecorded crashes (see Table E.1). To illustrate this variability, Blanco et al. (2016) [3.1] deliberately picked particular injury and PDO unrecorded percentages from different studies to create low, moderate, and high unrecorded crash estimates. The VTTI investigators added those to the police-recorded crash counts in the different SHRP2 regions (Blanco et al., 2016, their Chapter 3 and Appendix C) [3.1]. Blanco et al. (2016, Figure 2) [3.1] showed that it was possible to "bracket" the Google AV crash rate estimate with the 2013 U.S. national crash rate estimate, depending upon the specifically selected percentages of unrecorded injury and unrecorded PDO U.S. national crashes. In other words, Blanco et al. (2016, Figure 2) [3.1] demonstrated that AV crash rate estimates can be made to appear higher, lower, or about the same as the 2013 U.S. national average CV crash rate depending upon an arbitrary selection of the percentage of unrecorded crashes. The AV crash IRR estimates substantially change simply by the percentages of unrecorded injury and unrecorded PDO CV crashes selected from various survey studies (e.g., see Table E.1). More specifically, Blanco et al. (2016, Figure 2) [3.1] show that the Google AV Severity Levels 1 and 2 crash rate estimates can be made to appear arbitrarily crash causative, about the same, or crash preventive compared to the total U.S. national average crash rate estimate. Specifically, the crash rate of Google AVs compared to the national CV crash rate was: higher when calculated with a low estimate of unrecorded CV crashes; equal with a moderate estimate of unrecorded CV crashes; and lower with a high estimate of unrecorded CV crashes.

3.5.2. Discussion of Blanco et al. (2016) Unrecorded Crash Results

The results of Blanco et al. (2016, Figure 2) [3.1], as described in Section 3.5.1, demonstrate that comparing the Google AV crash rate estimate to the 2013 U.S. national CV crash rate estimate is arbitrary because it depends entirely upon what percentage of estimated unrecorded crashes is selected from a large set of choices (for example, see Table E.1).

Blanco et al. (2016, Figure 2) [3.1] further show a wide variation in police-recorded crash rates in Santa Clara County, CA, LA County, CA, and Kings County, WA. This site variation in unrecorded crashes means that direct comparisons of the Google AV crash rates in Mountain View and vicinity to police-recorded CV crashes are subject to site selection bias as well as unrecorded crash selection bias.

Furthermore, adjusting for site selection bias is further complicated because the effect of the bias may well be different for the different crash Severity Levels. For example, Figure 3.3B in Section 3.3.1 and Figure 3.4B in Section 3.3.2 show that the CV crash rate estimate is different in Mountain View than the SHRP2 sites in ways dependent upon the crash Severity Level. In particular, Figure 3.3B indicates that the Mountain View CV police-recorded crash rate is above the SHRP2 sites for CV injury crashes, but Figure 3.4B shows it is below the SHRP2 sites for Major PDO crashes. It is unknown and indeterminable from the limited available public information whether these severity differences are simply because of different reporting standards for the police jurisdictions in Mountain View compared to the SHRP2 sites. Or, are they valid differences in actual CV crash rates between the Mountain View and SHRP2 sites at different crash Severity Levels? Thus, methods that can reduce or eliminate such site selection bias and its interaction with crash Severity Levels are needed. Only then can a valid comparison be made between AV crash rates to CV crash rates at sites other than the AV ODD.

References

3.1. Blanco, M., Atwood, J., Russell, S., Trimble, T. et al., "Automated Vehicle Crash Rate Comparison Using Naturalistic Data," Virginia Tech Transportation Institute, January 2016, https://vtechworks.lib.vt.edu/handle/10919/64420, accessed May 19, 2021.

3.2. Schoettle, B. and Sivak, M., "A Preliminary Analysis of Real-World Crashes Involving Self-Driving Vehicles," Technical Report No. UMTRI-2015-34, The University of Michigan Transportation Research Institute, Ann Arbor, MI, 2015, http://www.umich.edu/~umtriswt/PDF/UMTRI-2015-34.pdf, accessed May 19, 2021.

3.3. SAE, "Taxonomy and Definitions for Terms Related to Driving Automation Systems for On-Road Motor Vehicles," SAE J3016_201806, SAE International, Warrendale, PA, 2018, https://www.sae.org/standards/content/j3016_201806/, accessed May 19, 2021.

3.4. StatsDirect, Statistical Software, Version 3.2.8, September 16, 2019, https://www.statsdirect.com/, accessed September 22, 2021.

3.5. SWITRS, "California Highway Patrol Statewide Integrated Traffic Records System," https://iswitrs.chp.ca.gov/Reports/jsp/index.jsp, accessed May 19, 2021.

3.6. HPMS, "California Department of Transportation Highway Performance Monitoring System Annual Reports 2009–2014," https://dot.ca.gov/programs/research-innovation-system-information/highway-performance-monitoring-system, accessed May 19, 2021.

3.7. Porta, M., *A Dictionary of Epidemiology*, 6th ed. (New York: Oxford University Press, 2014), 343 pp., http://irea.ir/files/site1/pages/dictionary.pdf, accessed May 19, 2021.

3.8. Rothman, K.J., *Epidemiology: An Introduction*, 2nd ed. (New York: Oxford University Press; 2012).

3.9. Rothman, K., Greenland, S., and Lash, T., *Modern Epidemiology*, 3rd ed. (Philadelphia, PA: Lippincott Williams & Wilkins, 2008).

3.10. NHTSA, "Traffic Safety Facts: 2013 Data," National Highway Traffic Safety Administration, Washington, DC, 2015, https://crashstats.nhtsa.dot.gov/Api/Public/ViewPublication/812169, accessed May 19, 2021.

4

Crash Selection Bias and Site Selection Bias: Dixit et al. (2016)

Contents

Note: Refer to Appendix B on page 149 for an explanation of the abbreviations and the definitions of unique terms and phrases.

Dixit et al. (2016) [4.1] are researchers at the Research Centre for Integrated Transport Innovation at the University of New South Wales.

4.1. Summary of Dixit et al. (2016)

Dixit et al. (2016) [4.1] investigated mainly AV disengagements (i.e., transitions from autonomous mode to manual mode). Disengagements are outside the direct scope of this book, although they were briefly mentioned in Chapter 2 (Sections 2.4.2.2.4 and 2.5) regarding their possible interactions with AV crashes. This book mainly reviews studies that estimated AV crash rates and compared them to various CV crash rates.

However, Dixit et al. (2016, pp. 6–9) [4.1] did a limited analysis of AV crashes during 14 months of CA testing from September 1, 2014, to November 30, 2015. This study period extended two months past the end date of September 30, 2015, of Schoettle and Sivak's (2015) [4.2] study analyzed in Chapter 2. It also extended one month past the end date of October 31, 2015, of Blanco et al.'s (2016) [4.3] study, analyzed in Chapter 3.

Table 1 in Dixit et al. (2016) [4.1] listed 12 AV crashes from the Google Reports [4.4]. Ten of those 12 crashes occurred during Dixit et al.'s [4.1] analysis period. On October 14, 2014, the first crash was from Delphi, which Dixit and colleagues did not further analyze. Nine of the subsequent AV crashes were from Google. Two of these nine Google AV crashes (February 26, 2015, and January 8, 2016) were transition crashes (Appendix C). Consistent with the two prior studies reviewed in Chapters 2 and 3, Dixit et al. (2016) [4.1] classified these two transition crashes in the same group as autonomous mode crashes (and not manual mode crashes) because "the autonomous mode was manually disengaged few moments prior to the accident" (Dixit et al., 2016, Table 1) [4.1]. Therefore, Dixit et al. (2016) [4.1] analyzed a total of nine Google AV crashes, from February 26, 2015, to February 14, 2016.

Eight of the nine Google AV crashes analyzed by Dixit et al. (2016) [4.1] (those from February 26, 2015, to August 20, 2015) overlapped eight of the 11 crashes analyzed by Schoettle and Sivak (2015) [4.2]. In Figure 2.1, these overlapping eight crashes are those on the right-hand side of the graph in 2015. Furthermore, these same eight 2015 crashes also coincided with eight of the 11 crashes analyzed by Blanco et al. (2016) [4.3]. The only crash analyzed by Dixit et al. (2016) [4.1] that did not coincide with the two prior studies was the last crash analyzed by Dixit et al. (2016) [4.1] on February 14, 2016.

Using these nine Google AV crashes, Dixit et al. (2016, p. 8) [4.1] estimated a crash frequency of one Google AV crash every 47,148 miles (0.047148 MVMT). Dixit et al. (2016) [4.1] did not specify in their article what total AV mileage they used to estimate that AV crash frequency. However, with nine crashes at a frequency of 1 per 47,148 miles, the AV mileage = $9*47{,}148/10^6 = 0.42433$ MVMT.

For the CV comparison group, Dixit et al. (2016) [4.1] selected the state of CA in its entirety, obtaining the police-recorded CV crash count from the CHP SWITRS database [4.5] for 2014. Dixit et al. (2016, p. 7) [4.1] estimated a 2014 CA crash frequency of one police-recorded CV crash per 2.07 MVMT. However, Dixit et al. (2016) [4.1] did not specify the 2014 police-recorded statewide CV crash count or the total 2014 statewide CV mileage that they used to estimate that CV crash frequency.

Dixit et al. (2016) [4.1] divided the AV crash frequency of 1 per 0.047148 MVMT by the statewide CV crash frequency of 1 per 2.07 MVMT to estimate a substantial 44 times increase in the Google AV crash frequency in Mountain View and vicinity from September 1, 2014, to November 30, 2015. Again, this substantially elevated ratio occurred when Google AV crash frequency in the Mountain View area was compared to the CV crash frequency in the state of CA for 2014.

4.2. Critical Analysis of Dixit et al. (2016)

It is of interest to estimate the confidence interval (CI) for Dixit et al.'s (2016) [4.1] substantial 44 times increase in the Google AV crash frequency compared to the statewide CV crash frequency. However, a CI for a crash frequency cannot be estimated from a crash frequency alone using epidemiologic methods; the crash count and mileage must also be known. Unfortunately, Dixit et al. (2016) [4.1] did not provide the AV mileage to estimate the AV crash frequency. They also did not provide the CV crash count or CV mileage used for their 2014 statewide CV crash frequency estimate.

Upon email request, R. Dixit and S. Chand [4.6] kindly provided the crash and mileage data that they used for their AV and CV crash frequency estimates in Dixit et al. (2016) [4.1]. Table 4.1 shows the data they provided in the 2×2 data matrix. Below the 2×2 data matrix are the estimates of the AV and CV crash rates (the reciprocal of Dixit et al.'s (2016)

TABLE 4.1 Dixit et al. (2016): Original data.

	Google AV[a]	2014 CA CV[e]
Crashes	**9**[b]	162,711[f]
MVMT	**0.42433**[c]	336,422[g]
Crash rate	21.2099[d]	0.4837[h]
IRR (95% CI)[i]	43.9 (21.4-80.5), P = 1.5E−12	

Replication of the data underlying Dixit et al.'s (2016) [4.1] AV and CV crash frequency estimates. Google AV total crash rate versus the 2014 CA police-recorded CV injury crash rate.

A **bold** number in the 2×2 data matrix indicates no identified error or bias.

Underlined words in the notes below are revised in later tables.

Notes:

[a] Google AV data from September 1, 2014, to November 30, 2015.

[b] Google AV total crash count from Dixit et al. (2016, Table 1) [4.1] from September 1, 2014, to November 30, 2015. Verified by a reanalysis using Google Reports [4.4] and the CA DMV [4.7] AV crash database.

[c] Google AV total autonomous mileage from Dixit and Chand (personal communication) [4.6]. Verified by a reanalysis using Google disengagement reports [4.8].

[d] The reciprocal of the Google AV crash rate estimate is one Google AV crash per 47,148 miles, replicating Dixit et al.'s (2016, p. 8) [4.1] AV crash frequency estimate, as expected.

[e] Statewide CV data for 2014.

[f] Total police-recorded crashes in CA in 2014 (Dixit and Chand, personal communication) [4.6]. The source provided by the authors was https://www.wmlawyers.com/2017/02/california-car-accident-statistics/, accessed May 26, 2021. This source at a legal firm states that these crashes were injury crashes only. That was verified by a reanalysis using the method described in this endnote: [4.9]. Thus, these crashes are <u>injury crashes only, with no fatal or PDO crashes</u>. The replication also found that these crashes were on all roads in CA, including city streets. However, Dixit et al. (2016, p. 8) [4.1] state, "The CHP only reports crashes on State, US, and interstate roads." While that statement is true, the investigators Dixit et al. (2016) [4.1] may not have realized that the CA SWITRS [4.5] crash database, which, although maintained by the CHP, also includes crashes on city streets and other roads under the jurisdiction of local police departments. A further explanation is given in this endnote: [4.10].

[g] 2014 CA MVMT from Dixit and Chand (personal communication) [4.6], from CHP. The replication determined that the CV mileage estimate included all CA city streets, roads, highways, and interstates (FHWA, 2015, Table 5) [4.11]. The 2014 CV MVMT estimate from Dixit and Chand (personal communication) [4.6] was replicated with the 2014 monthly mileage estimates in FHWA (2015, Table 5) [4.11].

[h] The reciprocal of this 2014 statewide CV crash rate estimate of <u>0.4837</u> CV crashes per MVMT is one CV crash per <u>2.07</u> MVMT, <u>replicating Dixit et al.'s (2016, p. 7) [4.1] CV crash frequency estimate, as expected</u>.

[i] The IRR, exact mid-P confidence levels, and two-tailed P were estimated with the StatsDirect rate ratio program [4.12].

[4.1] AV and CV crash frequencies). Thus, the replication of Dixit et al.'s (2016) [4.1] original data in the 2×2 data matrix in Table 4.1 now allows the estimation of the AV crash IRR and its confidence limits using the standard epidemiologic method in Table 1.1 as described in Sections 1.1 and 1.2 in Chapter 1.

Table 4.1 (left data column) provides the Google AV crash count of nine (Table 4.1, note b) and the Google AV mileage of 0.42433 (Table 4.1, note c) during the study period of Dixit et al. (2016) [4.1]. That Google AV crash count of nine (Table 4.1, note b) is based on the nine Google AV crashes listed in Dixit et al. (2016, Table 1) [4.1] from September 2014 to November 2015. The final Google AV crash count excludes the following three crashes listed in Table 1 in Dixit et al. (2016) [4.1]: the Delphi crash on October 14, 2014;

the Cruise Automation crash on January 8, 2016; and the Google AV crash on February 2, 2016.

The total Google AV MVMT of 0.42433 (Table 4.1, note c) was tabulated by Dixit and Chand (personal communication) [4.6] from the Google Reports [4.4], accumulated for all months from September 2014 to November 2015.

The Google AV total crash count divided by the Google AV total mileage estimates a Google AV total crash rate of 21.2099 crashes per MVMT (Table 4.1, note d). The reciprocal of that estimated AV total crash rate exactly replicates Dixit et al.'s (2016) [4.1, p. 8] Google AV crash frequency in Mountain View and vicinity of one AV crash "every 47,148" AV miles, as expected.

Table 4.1 (right data column) provides the 2014 CA police-recorded CV crashes of 162,711 (Table 4.1, note f) and the 2014 total CV mileage estimate on all CA roads of 336,422 MVMT (Table 4.1, note g).

The CV crash count divided by the CV mileage estimates a 2014 statewide CV total crash rate of 0.4837 crashes per MVMT (Table 4.1, note h). The reciprocal of that CV crash rate exactly replicates Dixit et al.'s (2016) [4.1, p. 7] 2014 statewide CV crash frequency of one CV crash "every 2.07 million miles" driven in CA, as expected.

4.2.1. Injury-Only CV Crash Systematic Error: (Table 4.1, Note f)

Dixit et al. (2016) [4.1] included only CA police-recorded injury CV crashes in Table 4.1 (note f), leaving out police-recorded fatal and PDO crashes. This systematic error in crash selection caused a substantial upward bias in the Google AV IRR estimate of 43.9 (Table 4.1, note i). This substantial systematic error occurred because the AV crashes (Table 4.1, note b) included all crash severity levels (fatal, injury, and PDO). In contrast, the CV crashes (Table 4.1, note f) included only CV injury crashes. In short, there was a substantial systematic error from leaving out fatal and PDO crashes from the 2014 statewide CV crash count. This error caused a substantial upward bias in the AV IRR estimate of 43.9 in Table 4.1 (note i).

Note that this method of counting only CV injury crashes could also have been classified as a crash selection bias because the method uses different criteria for selecting CV crashes (injury-only crashes) than AV crashes (all crashes). However, it is more plausible that this issue was an inadvertent error by the investigators in not realizing that their source (a CA legal firm) for the statewide CV crash count (see Table 4.1, note f) included only injury crashes. So this issue was classified as a substantial systematic error rather than a crash selection bias.

Correcting the Injury-Only CV Crash Error by Adding Fatal and PDO Crashes

In the 2014 statewide CV crash count, the substantial systematic error was corrected by adding fatal and police-reported PDO CV crashes to the CV injury crash count in Table 4.1 (note f). As a result, the CV crash selection criteria now correctly matched the AV selection criteria for crash severity levels after removing this substantial systematic error. This correction to the statewide CV total crash count by itself caused a substantial 43.8% decrease in the AV IRR (to 30.5) versus the original AV IRR (43.9) (Table 4.1, note i).

But the CV crash count estimate in Table 4.1 (note f) still had unrecorded crashes that were not counted. But the AV estimate counted all crashes (Table 4.1, note b). Using different criteria for counting AV and CV crashes means that unrecorded crash selection bias remains in the AV IRR estimate in Table 4.1, as described in the next section.

4.2.2. Unrecorded Crash Selection Bias: (Table 4.1, Notes b and f)

A substantial unrecorded crash selection bias in the Dixit et al. (2016) [4.1] original data in Table 4.1 was that the AV crash count (Table 4.1, note b) included all crashes (both police-recorded and unrecorded). But the CV crash count (Table 4.1, note f) included only police-recorded crashes. This unrecorded crash selection bias could not be unambiguously removed from Table 4.1 for reasons explained in Section 4.2.5.

4.2.3. Improving Accuracy: Statewide CV MVMT (Table 4.1, Note g)

Dixit et al.'s (2016) [4.1] 2014 statewide CV mileage total estimate of 336,422 (Table 4.1, note g) was based on the FHWA mileage database [4.11]. The NHTSA (2014) [4.13] estimate for 2014 CA CVs was lower, at 332,857 MVMT. However, it is plausible that the CA HPMS databases [4.14, 4.15] have more accurate mileage estimates than the FHWA and NHTSA federal databases [4.11, 4.13] for many reasons. One reason is that the national databases may pull the state estimates before they are final. The final state HPMS (2014) [4.14] estimate is 333,965 MVMT, in between the FHWA [4.11] and NHTSA [4.13] CA mileage estimates. Therefore, the reanalysis updated the statewide CV mileage Federal estimate to the 2014 state-level HPMS estimate [4.15]. This minor adjustment to the CA MVMT caused a negligible 0.7% decrease in the AV IRR to 43.5 compared to the original IRR of 43.9 (Table 4.1, note i).

4.2.4. Partly-Corrected Dixit et al. (2016) Data (Table 4.2)

Table 4.2 gives the partly-corrected Dixit et al. (2016) [4.1] data based on the corrections in Sections 4.2.1 and 4.2.3.

Table 4.2 (note f) eliminates the injury-only CV systematic error (Section 4.2.1). Table 4.2 does so by adding police-recorded CV fatal and PDO crashes to the police-recorded CV

TABLE 4.2 Dixit et al. (2016): Partly-corrected CV data.

	Google AV[a]	2014 CA CV[e]
Crashes	**9**[b]	**233,941**[f]
MVMT	**0.424331**[c]	**333,965**[g]
Crash rate	21.2[d]	0.700[h]
IRR (95% CI)[i]	30.3 (14.8–55.6), P = 3.7E–11	

Google AV crash rate versus partly-corrected CA police-recorded CV total crash rate.

A **bold** number in the 2×2 table indicates no identified error or bias.

Notes:

[a–e] Same as Table 4.1.

[f] Total 2014 police-recorded fatal + injury + PDO CV crashes in CA (SWITRS database [4.5]).

[g] Same as Table 4.1, except the mileage is from the CA 2014 HPMS [4.14, 4.15] mileage database.

[h] The corrected 2014 statewide CV total crash rate estimate. The reciprocal estimates a statewide CV total crash frequency of one CV crash per 1.43 MVMT.

[i] Same as Table 4.1.

injury crashes. In other words, Table 4.2 (note f) updates the 2014 statewide CV crash total to include all police-recorded crashes, not just injury crashes, as per the correction method described in Section 4.2.1. However, there is still an unrecorded crash selection bias, as noted in Section 4.2.2. This issue is addressed in Section 4.2.5.

Table 4.2 (note g) also shows the new 2014 statewide CV total mileage estimate (Section 4.2.3), revised to improve accuracy. As a result of these revisions, the 2014 statewide CV total crash rate increased from 0.4837 (Table 4.1, note h) to 0.700 (Table 4.2, note h). This increase in the CV total crash rate estimate caused the Google AV crash IRR estimate to decrease from 43.9 (Table 4.1, note i) to 30.3 (Table 4.2, note i).

4.2.5. Unrecorded Crash Selection Bias: (Table 4.2, Notes b and f)

The partly-corrected Dixit et al. (2016) [4.1] Google AV crash IRR estimate of 30.3 (Table 4.2, note i) still contains a substantial remaining upward bias from the unrecorded crash selection bias previously described in Section 4.2.2 regarding Table 4.1. Specifically, there are different crash selection criteria for the AV and CV groups.

The Google AV crash group (Table 4.2, note b) was selected by Dixit et al. (2016) [4.1] to include <u>all</u> AV crashes (both police-recorded and police-unrecorded). In contrast, the statewide CV crash group selected by Dixit et al. (2016) [4.1] contains <u>only</u> police-recorded crashes (Table 4.2, note f). This criterion mismatch is an unrecorded crash selection bias that caused a substantial upward bias in the AV crash IRR estimate of 30.3 in Table 4.2 (note i).

The upward bias from this imbalance in the crash selection criteria could be adjusted by making assumptions about the percentage of unrecorded injury and PDO crashes (see Appendix E). This method adds estimates of unrecorded CV crashes to the recorded CV crashes, as per Schoettle and Sivak (2015) [4.2] (Chapter 2, Section 2.1.2) and Blanco et al. (2016) [4.3] (Chapter 3, Section 3.5.1).

However, any attempt to add unrecorded CV crashes to the police-recorded CV crashes contains the potential for additional bias because of the uncertainties in the percentages of unrecorded injury CV crashes and unrecorded PDO CV crashes (Appendix E). As Blanco et al. (2016, Figure 2) [4.3] correctly demonstrated, the AV IRR effect size estimate is entirely dependent upon arbitrary assumptions about the unknown number of unrecorded injury crashes and PDO crashes. Blanco et al. (2016, Figure 2) [4.3] showed that the AV crash rate could be made to appear arbitrarily higher, lower, or equal to the CV crash rate simply by using different estimates of the percentages of unrecorded injury and PDO crashes from various studies (see Chapter 3, Section 3.5).

Another way to adjust for unrecorded crash selection bias requires no assumptions about the number of unrecorded crashes. The method is to use only police-recorded AV crashes, as was done in the reanalysis of the data of Schoettle and Sivak (2015) [4.2] (Chapter 2, Section 2.4.1). The AV and CV groups would then have the same selection criterion (police-recorded crashes only), eliminating the unrecorded crash selection bias. However, this method gives rise to a zero AV crash rate because there were no police-recorded AV crashes for the Google AV crashes in the period of these studies (Chapter 2, Section 2.4.1, Table 2.5) at any crash severity. Surprisingly, not even the AV <u>injury</u> crashes had a police record (see questions raised in Appendix E about why AVs have few recorded crashes in police databases). Hence, the method of using only police-recorded AV crashes to adjust for unrecorded crash selection bias will not be pursued further for the Dixit et al.

[4.1] Google AV data, despite its success in the reanalysis of the Schoettle and Sivak (2016) [4.2] Google AV data (Chapter 2, Section 2.4.1 and Table 2.5). Consequently, the partly-corrected Schoettle and Sivak (2016) [4.2] Google AV crash IRR of 30.3 in Table 4.2 is invalid because of unrecorded crash selection bias. Unfortunately, a substantial site selection bias is also present, as shown in Section 4.2.6.

4.2.6. Site Selection Bias: AV and CV Crashes (Table 4.2, Notes b and f)

Section 4.2.5 showed that unrecorded crash selection bias could not be eliminated for the partly-corrected Dixit et al. (2016) [4.1] data in Table 4.2. However, even if that bias could somehow be eliminated, a remaining site selection bias would still be present in the IRR estimate of 30.3 (Table 4.2, note i). The CV crash rate was different between the AV and CV sites. Google AV test trials were conducted in Mountain View and vicinity. However, the city of Mountain View had a higher police-reported overall crash rate in 2014 than the state of CA. This fact causes a site selection bias, as illustrated by Table 4.3.

Table 4.3 shows that Mountain View had a higher CV crash rate than the state of CA in 2014. Because the Google AV tests were conducted mainly in Mountain View, comparing the AV crash rate to the statewide CV crash rate in Tables 4.1 and 4.2 upwardly biased the AV crash IRR because of the site selection bias demonstrated by the data and analysis in Table 4.3.

4.2.7. Discussion and Conclusion of Dixit et al. (2016) Review

Two systematic errors were identified in the Dixit et al. (2016) [4.1] data, each of which caused an upward bias in the IRR:

1. A minor systematic error in the statewide CV MVMT estimate (Section 4.2.3).
2. A substantial systematic error by using injury-only CV crashes (Section 4.2.1).

TABLE 4.3 Dixit et al. (2017): Site selection bias.

	2014 Mountain View CV[a]	2014 CA CV[e]
Crashes	**172**[b]	**233,941**[f]
MVMT	**209.26**[c]	**333,965**[g]
CV crash rate	0.822[d]	0.700[h]
IRR (95% CI)[i]	1.17 (1.0–1.36), $P = 0.04$	

2014 police-recorded CV crash rate in Mountain View versus 2014 revised statewide CV total crash rate.

A **bold** number in the 2×2 data matrix indicates no identified error or bias.

Notes:

[a] Mountain View CV data from January 1, 2014, to December 31, 2014.

[b] Total police-recorded crashes on Mountain View city streets in 2014 (fatal, injury, and PDO) (SWITRS) [4.5].

[c] MVMT on Mountain View city streets in 2014 (HPMS, 2014) [4.14].

[d] CV total crash rate Total in Mountain View city streets in 2014.

[e–i] Same as Table 4.2.

Two biases were identified in the Dixit et al. (2016) [4.1] data, each of which caused a substantial upward bias in the IRR:

1. Unrecorded crash selection bias (Sections 4.2.2 and 4.2.5).
2. Site selection bias (Section 4.2.6).

The two systematic errors were corrected. Unfortunately, the two biases in the Dixit et al. (2016) [4.1] could not be adjusted without making uncertain and arbitrary assumptions about unrecorded crashes (see Appendix E). Therefore, a corrected and unbiased final AV crash IRR was not estimated for Dixit et al. (2016)'s [4.1] data, as it had been in Chapter 2 (Section 2.3, Table 2.5) for Schoettle and Sivak's (2015) [4.2] data, and Chapter 3 (Section 3.3, Table 3.2) for Blanco et al.'s (2016) [4.3] data.

References and Endnotes

4.1. Dixit, V.V., Chand, S., and Nair, D.J., "Autonomous Vehicles: Disengagements, Accidents and Reaction Times," *PLOS ONE* 11, no. 12 (2016): e0168054, doi:10.1371/journal.pone.0168054, https://doi.org/10.1371/journal.pone.0168054, accessed May 19, 2021.

4.2. Schoettle, B. and Sivak, M., "A Preliminary Analysis of Real-World Crashes Involving Self-Driving Vehicles," Technical Report No. UMTRI-2015-34, The University of Michigan Transportation Research Institute, Ann Arbor, MI, 2015, http://www.umich.edu/~umtriswt/PDF/UMTRI-2015-34.pdf, accessed May 19, 2021.

4.3. Blanco, M., Atwood, J., Russell, S., Trimble, T. et al., "Automated Vehicle Crash Rate Comparison Using Naturalistic Data," Virginia Tech Transportation Institute, January 2016, https://vtechworks.lib.vt.edu/handle/10919/64420, accessed May 19, 2021.

4.4. Google Reports, "Google Self-Driving Car Project Monthly Reports," Mountain View, CA, May 2015–November 2016. Note: Monthly reports are longer publicly available from Google. However, 19 monthly Google Reports from May 2015 to November 2016 are available at https://github.com/wgetsnaps/ca-dmv-autonomous-vehicle-accidents/tree/master/google-pdfs, accessed May 19, 2021.

4.5. SWITRS, "California Highway Patrol Statewide Integrated Traffic Records System," https://iswitrs.chp.ca.gov/Reports/jsp/index.jsp, accessed May 19, 2021.

4.6. Dixit, V. and Chand, S., Personal communication, August 25, 2019.

4.7. CA DMV, "Report of Traffic Collision Involving an Autonomous Vehicle (OL 316)," California Department of Motor Vehicles, 2014–2019, https://www.dmv.ca.gov/portal/uploads/2021/05/OL316.pdf, accessed August 14, 2021.

4.8. Google, "Google Self-Driving Car Testing Report on Disengagements of Autonomous Mode December 2015," Cumulative Autonomous Mileage from 9/1/2014–11/30/2015, https://www.consumerwatchdog.org/resources/cadmvdisengagereport-dec.2015.pdf, accessed May 19, 2021.

4.9. Note: The California yearly total crash count is not available from SWITRS Report 1, "Collisions and Victims by Motor Vehicle Involved With," in the query system for the online California SWITRS database [4.5]. If a query request is made for "All Jurisdictions" and "All Locations," the error is received: "Cannot select State wide report for SWITRS Report." Therefore the raw data was requested for all of CA for 2014. A count in Excel was then made from the total fatal, injury, and PDO crashes from the raw data file.

4.10. Note: A party may report the crash to a local police department with jurisdiction for the crash location. But the police may or may not fill out a formal *Police Accident Report* (*PAR*). If the local police department chooses to fill out a PAR, they may then decide to file or not to file a formal crash record with the CA CHP for inclusion in the CA SWITRS [4.5] database (see Appendix E). As previously noted in Table 4.1 (note f), a separate minor issue was that Dixit et al. (2016, p. 8) [4.1] misstated that the SWITRS database [4.5] contains only "crashes on State, U.S., and interstate roads." The SWITRS database [4.5] includes all crashes recorded to the state police database by all police jurisdictions in CA, including crashes on city streets, as reported by all the police departments in every city and county in California. The SWITRS database [4.5] also contains crashes on rural roads and private property. The SWITRS [4.5] query system has options that allow the selection of crashes on city streets, state and interstate highways, or rural roads and private property. Fortunately, this minor misstatement by Dixit et al. (2016, p. 8) [4.1] did not affect any of their initial results in Table 4.1 or their corrected and reanalyzed data in Table 4.2 and 4.3.

4.11. FHWA, "Table 5. Travel Monitoring: Traffic Volume Trends," 2015, https://www.fhwa.dot.gov/policyinformation/travel_monitoring/tvt.cfm, accessed May 19, 2021.

4.12. StatsDirect, Statistical Software, Version 3.2.8, September 16, 2019, https://www.statsdirect.com/, accessed September 22, 2021.

4.13. NHTSA, "Select on the map below to see a State report," 2014, https://cdan.nhtsa.gov/STSI.htm#, accessed May 19, 2021. Note: Bring up the Federal Map and then click on CA.

4.14. HPMS, "California Department of Transportation Highway Performance Monitoring System Annual Reports 2009–2014," https://dot.ca.gov/programs/research-innovation-system-information/highway-performance-monitoring-system, accessed May 19, 2021.

4.15. HPMS, "Road Data from Highway Performance Monitoring System," 2014, https://dot.ca.gov/-/media/dot-media/programs/research-innovation-system-information/documents/california-public-road-data/prd20141-a1ly.pdf, accessed May 19, 2021. Note: The bottom row of Table 6 across all CA jurisdictions shows a total 2014 mileage estimate of 914,971.93 Daily Vehicle Miles Traveled x 1000. Multiply by 365 and divide by 1,000 to obtain the 2014 CA statewide MVMT estimate of 333,965 in Table 4.2, note g.

5

Crash Selection Biases and Site Selection Biases: Teoh and Kidd (2017)

Contents

Note: Refer to Appendix B on page 149 for an explanation of the abbreviations and the definitions of unique terms and phrases.

Teoh and Kidd (2017) [5.1] analyzed Google AV crashes and mileage in public road tests from 2009 to 2016. They compared the AV crashes and mileage primarily to Mountain View CV crashes and mileage. The AV tests were mainly in Mountain View, CA, which had the vast majority of the Google AV crashes, and likely most of the AV test mileage. Teoh and Kidd (2017) [5.1] divided their data analysis into two sets: Phase 1 (2009–12/31/15) called "Mountain View," and combined Phases 1 and 2 (2009–August 21, 2016) called "All States." The second phase includes data from other states.

5.1. Phase 1 "Mountain View": 2009–December 31, 2015

The Phase 1 analysis of Teoh and Kidd (2017) [5.1] used data predominately from Mountain View.

5.1.1. Teoh and Kidd's (2017) Phase 1 Original Data

5.1.1.1. AV Crashes

Teoh and Kidd (2017) [5.1] tabulated Google AV crash data from 2009 to December 31, 2015, using CA DMV "OL-316" forms [5.2] and monthly Google Reports (2015–2016) [5.3]. Teoh and Kidd (2017, Table 2) [5.1] counted ten crashes in autonomous mode.

Unlike the three prior studies (see Chapters 2, 3, and 4), Teoh and Kidd (2017) [5.1] did not count two transition crashes from autonomous mode to manual mode as in the same group as autonomous crashes for analyzing crash rates.

Teoh and Kidd (2017) [5.1] judged three of the ten autonomous mode AV crashes as "police-reportable." By "police-reportable," Teoh and Kidd (2017, p. 58) [5.1] meant AV crashes that they "judged likely to be reportable to police." That is, those are crashes they "deemed" serious enough to be likely reported to the police. However, Teoh and Kidd (2017, p. 10) [5.1] acknowledge, "It was unknown if police crash reports were filed." For AVs, Teoh and Kidd (2017) [5.1] counted the number of crashes, not the number of vehicles involved in the crashes. For example, a multi-vehicle crash involving one AV and one or more CVs were counted as one AV crash.

5.1.1.2. AV Mileage

For AV mileage during Phase 1, Teoh and Kidd (2017) [5.1] cited Google's (2015, December) [5.3] report, which showed a total of 1,372,111 miles in autonomous mode. Dividing the three AV crashes they deemed "police-reportable" by the autonomous driving miles of 1,372,111 and multiplying by a million gave rise to Teoh and Kidd's (2017, Table 2) [5.1] Phase 1 AV crash rate estimate of 2.19 "police-reportable" AV crashes per MVMT.

5.1.1.3. CV Crash-Involvement

To estimate the number of "CV-involved" crashes during Phase 1, Teoh and Kidd (2017) [5.1] counted crash-involved CVs at four different geographic levels: City (Mountain View), County (Santa Clara), State (CA), and National (USA). However, the investigators devoted most of their analysis, discussion, and conclusions to the CV data in Mountain View (City level). These CV data are what are reviewed and analyzed in this Chapter. Teoh and Kidd (2017) [5.1] selected CV crashes in Mountain View using three criteria.

Three CV Crash-Involvement Selection Criteria for Mountain View, CA

1. Count the number of passenger vehicles involved in a crash (vehicle level). Teoh and Kidd (2017) [5.1] did not count the number of crashes in which one or more passenger vehicles were involved (crash level). The vehicle level count is always larger than the crash level count because more than one vehicle is usually involved in a crash.
2. Do not count motorcycles, motor scooters, trucks, tractor-trailers, bicycles, etc., involved in a crash.
3. Count crash-involved passenger vehicles on all roads, including state routes, highways, and freeways passing through Mountain View city limits, not just crashes on city streets alone.

Using these three CV selection criteria, Teoh and Kidd (2017, Table 3) [5.1] tabulated 9,198 passenger CVs involved in police-recorded crashes on all roads within Mountain View city limits from 2009 to 2015. Appendix I replicates their 9,198 passenger vehicle crash-involvement count using these three criteria.

5.1.1.4. CV Mileage

To estimate CV mileage in Mountain View during their Phase 1, Teoh and Kidd (2017) [5.1] used three mileage selection criteria:

CV Mileage Selection Criteria for Mountain View, CA

1. Mileage for <u>all</u> CV types (passenger vehicles, motorcycles/scooter, truck, tractor-trailer, etc.), not just for passenger vehicles. (Note that this criterion for mileage is broader than for CV crash involvement in Criterion 2 in Section 5.1.1.3 above.)
2. Mileage <u>only on city streets</u>, not on highways, freeways, and state routes that passed through Mountain View city limits. The HPMS dataset [5.4] they used gives the miles traveled only on Mountain View city streets, not on other roads within the Mountain View city limits.
3. Use 2014 CV mileage to estimate 2015 CV mileage. Unfortunately, the 2015 CV mileage was not available in the HPMS dataset [5.4] at the time of Teoh and Kidd's (2017) [5.1] study.

With these three mileage selection criteria, Teoh and Kidd (2017) [5.1] estimated the number of CV miles traveled in Mountain View from 2009 to 2015 as 1,517,930,400, using mileage data from the HPMS database [5.4].

To estimate the passenger vehicle "crash-involvement rate", Teoh and Kidd (2017, Table 3) [5.1] took the total number of passenger vehicles "involved" in a crash (given in Section 5.1.1.3 as 9,198), divided by the total CV driving miles (provided in Section 5.1.1.4 as 1,517,930,400), and then multiplied by a million. Their final CV "crash-involvement" rate estimate for Mountain View in Phase 1 was, therefore, 6.06 passenger vehicles involved in crashes per MVMT. Section 5.1.1.5 summarizes their original data.

5.1.1.5. Summary of Teoh and Kidd's (2017) Phase 1 Original Data

Table 5.1 summarizes the original data on which Teoh and Kidd (2017) [5.1] based their results and conclusions for Phase 1.

Teoh and Kidd's (2017) [5.1] AV crash rate estimate of 2.19 (Table 5.1, note d) is smaller than their passenger vehicle crash-involvement rate estimate of 6.06 (Table 5.1, note h). Hence, Teoh and Kidd (2017) [5.1] conclude that this difference indicates that "Google cars had a much lower rate of police-reportable crashes per million VMT than human drivers in Mountain View, Calif., during 2009–2015." They further conclude, "These results suggest Google self-driving cars, while a test program, are safer than conventional human-driven passenger vehicles."

Teoh and Kidd (2017) [5.1] used conventional statistical significance criteria to conclude that AVs were safer than CVs. However, statistical significance criteria are not used in this review (see Section 1.5). Instead, this review does an epidemiologic analysis of their Phase 1 data. Table 5.1 organizes the original Teoh and Kidd (2017) [5.1] Phase 1 data into the epidemiologic design for estimating an IRR (see Table 1.1 in Chapter 1).

The Google AV crash IRR is estimated from the Table 5.1 data by dividing the Google AV crash rate of 2.19 (Table 5.1, note d) by the Mountain View passenger vehicle

TABLE 5.1 Teoh and Kidd (2017): Original Phase 1 data.

	Google AV[a]	Mountain View CV[e]
Crashes	3[b]	9,198[f]
MVMT	1.372111[c]	1,517.93[g]
Rate (95% CI)[i]	2.19[d] (0.44–6.39)	6.06[h] (5.93–6.18)
IRR (95% CI)[j]	0.36 (0.092–0.982), P = 0.045	

Google AV crash rate in the Mountain View area and Austin, TX versus the "passenger vehicle crash involvement" rate in Mountain View from 2009 to 2015.

A **bold** number in the 2×2 table above means no identified error or bias.

The underlined footnote words in the notes below are revised in later tables.

Notes:

[a] Cumulative Google AV data in the Mountain View area, 2009–December 31, 2015, and Austin, TX from July 1, 2015, to December 31, 2015 (Teoh and Kidd, 2017, Table 2) [5.1].

[b] Google AV crashes judged "police-reportable" by Teoh and Kidd (2017, Table 2) [5.1]. (Two transition crashes were excluded.)

[c] Total Google AV mileage (Teoh and Kidd, 2017, Table 2) [5.1]. Cumulative mileage from Google Reports (2015, December) in Mountain View, CA, and vicinity, plus 0.280673 MVMT for Google AV mileage in Austin, TX from July 1, 2015, to December 31, 2015.

[d] Google AV crash rate estimate for Phase 1 (Teoh and Kidd, 2017, Table 2) [5.1].

[e] Cumulative CV data in Mountain View, CA, from 2009 to 2015 (Teoh and Kidd, 2017, Table 3) [5.1].

[f] Passenger vehicles involved in police-recorded crashes within Mountain View city limits on all roads, including state roads, highways, and freeways passing through Mountain View. (Teoh and Kidd, 2017, Table 3) [5.1]. Source: SWITRS (2009–2015) [5.5]. The "Party Record" for a crash lists all involved vehicles in a crash.

[g] CV mileage (all vehicle types) on Mountain View city streets only, 2009–2015; 2014 mileage repeated for 2015 (Teoh and Kidd, 2017, Table 3) [5.1]. Source: HPMS (2009–2014) [5.4].

[h] Passenger vehicle crash-involvement rate in police-recorded crashes on all roads in Mountain View city limits (Teoh and Kidd, 2017, Table 3) [5.1].

[i] AV crash rate and CV crash-involvement rate and their CIs as estimated by Teoh and Kidd (2017, Tables 2 and 3) [5.1].

[j] The IRR, exact mid-P confidence levels, and two-tailed P were estimated with the StatsDirect rate ratio program [5.6]. Note: The IRR, CI, and P-value were estimated in the current review, not by Teoh and Kidd (2017) [5.1].

"involvement" rate of 6.06 (Table 5.1, note h). The resulting Google AV IRR estimate is 0.36, with an exact mid-P 95% CI of (0.092–0.982) and with P-value = 0.045.

This Google AV IRR estimate of 0.36 is below one, which indicates a crash-reducing or "protective" effect of AV exposure in the absence of bias. (An IRR estimate above one would show a crash-increasing or "causative" effect of AV exposure; see Section 1.1.) Assume the Teoh and Kidd (2017) [5.1] original Phase 1 data in Table 5.1 are free from error and bias. Then the AV crash IRR estimate of 0.36 indicates that the Google AV fleet resulted in about a 64% decrease in the "police-reportable" AV crash rate versus the passenger vehicle involvement rate for "police-recorded" crashes. Thus, AVs' 64% crash reduction in Teoh and Kidd's (2017) [5.1] original data in Table 5.1 was confirmed using epidemiologic analysis methods. This confirmation appears to provide support for Teoh and Kidd's (2017) [5.1] conclusion that prototype Google AVs "are safer than conventional human-driven passenger vehicles." The 95% confidence limits for the IRR estimate suggest that a Google AV "police-reportable" crash rate ranging from a 91% decrease to a 2% decrease versus the "passenger vehicle involvement rate" in police-recorded crashes is also reasonably compatible with Teoh and Kidd's (2017) [5.1] original data in Table 5.1.

However, this protective AV crash IRR estimate of 0.36 based on the Teoh and Kidd (2017) [5.1] data in Table 5.1 is invalid. It should not be cited because all four estimates in the 2×2 cells in Table 5.1 contain systematic errors and biases, as shown in Section 5.1.2.

5.1.2. Critical Analysis of Teoh and Kidd (2017): Phase 1

All four 2×2 cells in the original Teoh and Kidd (2017) [5.1] Table 5.1 (notes b, c, f, and g) contain systematic errors or biases that were unrecognized by Teoh and Kidd (2017) [5.1]. The number of "police-reportable" AV crashes (Table 5.1, note b) was a judgment by the investigators and is uncertain compared to actual police reports, not to mention actual police records filed in state databases. The AV and CV mileage estimates (Table 5.1, notes c and g) have a site selection bias. The geographic sites are not matched between the Google AV sites (Mountain View and vicinity plus Austin, TX) and the CV exposure sites (Mountain View, CA, and vicinity). In addition, the count of the passenger vehicles involved in crashes (Table 5.1, note f) has both a crash selection bias and a systematic error. Consequently, the crash-reducing AV IRR of 0.36 estimated from the Teoh and Kidd (2017) [5.1] original data in Table 5.1 is invalid. The errors and biases in the Teoh and Kidd (2017) [5.1] Table 5.1 data are documented in Sections 5.1.2.1, 5.1.2.2, 5.1.2.3, 5.1.2.4, 5.1.2.5, 5.1.2.6, 5.1.2.7, 5.1.2.8, 5.1.2.9, and 5.1.2.10, with corrections and adjustments where possible.

5.1.2.1. AV Crash Selection Bias 1: "Police-Reportable" AV Crashes (Table 5.1, Note b)

It is common knowledge that not all crashes have a formal police report written for them (see Appendix E). For example, in many crashes, the police are never contacted. In this instance, the police obviously cannot write up a Police Accident Report. Assume the police are contacted. In some jurisdictions, the police may not come to the crash scene, particularly if the crash does not involve an injury. Again, the police could not write up a Police Accident Report if they haven't been to the crash scene. If the police are called and come to the crash scene, the police may still decide not to write up a Police Accident Report. They may make only a "desk report." Even Police Accident Reports may not always be recorded in state crash databases. Therefore, very few crashes are ever recorded in a database accessible to outside safety researchers, except for crashes involving a fatality (see Appendix E).

On the contrary, by the CA regulations [5.7] that permit testing of AVs with a test driver on CA public roads, <u>every</u> AV crash, no matter how minor, is required to be documented and posted to the CA DMV's AV crash website on an OL-316 AV crash report form [5.2]. A direct comparison of <u>all</u> AV crashes to <u>only</u> police-recorded CV crashes would, therefore, lead to the biased conclusion that AVs have substantially higher crash rates than CVs. The overestimate occurs because of crash selection bias. The exposed (AV) and unexposed (human-driven CV) crash groups are selected using different crash selection criteria: <u>all</u> crashes versus police-recorded crashes, respectively.

Teoh and Kidd (2017) [5.1] correctly recognized this unreported crash selection bias and attempted to reduce it. They used a different method than did Schoettle and Sivak (2015) [5.8] (Section 2.1.2 in Chapter 2), who tried to estimate unreported CV crashes. Teoh and Kidd instead attempted to match the crash selection criterion for AVs to that for police-reported CV crashes. Teoh and Kidd attempted this match by deciding which AV crashes they "expected" would have a police record. Expert judgment has played an important and valuable historical role in many traffic safety issues (Hauer, 2019) [5.9]. However, it is plausible that the three AV crashes that Teoh and Kidd (2017) [5.1] judged "police-reportable"

(Table 5.1, note b) might not have had an actual police record (a false-positive error). This form of crash selection bias would upwardly bias Teoh and Kidd's (2017) [5.1] AV crash count in Table 5.1 (note b). Teoh and Kidd (2017, p. 10) [5.1] acknowledged, "It was unknown if police crash reports were filed." Teoh and Kidd (2017, p. 60) [5.1] admitted that their "police-reportable" metric overestimated actual police-reported crashes: "Google car crashes that were reportable to police likely overestimated the crash rate, relative to studying human drivers' crashes that were reported to police." Note that there was an additional bias unrecognized by Teoh and Kidd (2017) [5.1]. Simply "reporting" a crash to the police does not guarantee that the police will file a police record for the crash into a county or state crash database (see Appendix E).

This "police-reportable" AV crash bias is complicated by the fact that Teoh and Kidd (2017) [5.1] did not compare the rate of "police-reportable" AV crashes to the rate of police-reported CV crashes. Instead, the CV rate that Teoh and Kidd (2017) [5.1] estimated was the number of vehicles involved in a CV crash (see Section 5.1.2.5). Counting vehicles involved rather than crashes biases the CV rate estimate too high relative to the AV rate, which counted only crashes. In turn, biasing the CV rate too high would bias the AV IRR estimate too low, no matter how many AV crashes Teoh and Kidd (2017) [5.1] judged to be "police-reportable" (see Section 5.1.2.5).

Adjustment Method: Use Police-Recorded AV Crashes. One method to adjust for a potential "police-reportable" crash selection bias is to count only those AV crashes with actual police records, not just those crashes that Teoh and Kidd (2017) [5.1] judged were "likely" to have a police report. In other words, the AV crashes can be selected using the same criterion as the CV group—both must have an actual police record. A close inspection of the SWITRS [5.5] police-recorded crash database was independently undertaken by two researchers (the author and his senior research associate) to find any actual police-recorded AV crashes using the matching method in Section 1.4 in Chapter 1. The result was that none of the ten AV crashes in Phase 1 of the Teoh and Kidd (2017) [5.1] study matched a crash record in the CA SWITRS [5.5] crash database. This finding is consistent with the conclusion of Blanco et al. (2016, p. 9) [5.10] that "no PARs were filed for any of the Self-Driving Car-involved crashes for which law enforcement were on the scene." Thus, the AV crash rate and crash IRR became zero in Phase 1 after removing the unrecorded crash selection bias. That result suggests (in the absence of other errors and biases) that AVs are even safer than concluded by Teoh and Kidd (2017) [5.1].

However, this null result for Google AV crashes with an actual police record is surprising. During the Phase 1 Google AV test period, CA state law required that any crash with a fatality, injury, or more than $750 of property damage must be reported to the police. It is easy to achieve $750 of property damage on a Google AV from almost any collision because of its expensive external sensors. However, the following three serious AV crashes (likely the three crashes judged as "police-reportable" by Teoh and Kidd) had no SWITRS [5.5] police report: A Major PDO crash during March 2013, an injury crash on July 1, 2015, and a PDO crash on November 2, 2015, as follows.

1. <u>March 2013, a Major PDO two-vehicle crash on Highway 680S in San Jose</u>. The Google AV was going 63 mph in autonomous mode "when another vehicle traveling in the adjacent lane veered into the side of the Google AV." No injuries were reported at the scene, but the damage was judged by Blanco et al. (2016) [5.10] to meet the SHRP2 study's $1500 threshold for a Major PDO crash, which is double the CA $750 police reporting threshold for the CA DMV at the time of this crash

(https://legiscan.com/CA/text/SB438/id/1143271). This early crash was the only AV crash on a highway ever reported by Google. All subsequent Google AV crashes were on city streets, suggesting that Google subsequently geo-fenced its AV ODD to city streets.

2. July 1, 2015, a four-vehicle crash on a city street in Mountain View. See the description in Appendix E.4.1.
3. November 2, 2015, a two-vehicle crash on a city street in Mountain View. A rear-end crash into the Google AV at about four mph.

Why do these three crashes (or any other Google AV crash in Phase 1) have no police record in the SWITRS [5.5] database? Appendix E.4 provides some possible explanations.

In sum, examination of the CA SWITRS [5.5] crash database reveals that the number of police-recorded Google AV crashes (selected with the same criteria as CV police-recorded crashes in Mountain View or vicinity) is zero (no police report exists) rather than three (as estimated by Teoh and Kidd in Table 5.1, note b).

5.1.2.2. AV Crash Selection Bias 2: Censoring Transition Crashes (Table 5.1, Note b)

Teoh and Kidd (2017, p. 58) [5.1] counted crashes for Google cars "only while the vehicles were operated in autonomous mode." They did not count crashes during the transition from autonomous mode to manual mode, unlike the three prior studies: See Chapter 2, Section 2.1.1 for Schoettle and Sivak (2015) [5.8]; Chapter 3, Section 3.1.1 for Blanco et al. (2016) [5.10]; and Chapter 4, Section 4.1 for Dixit et al. (2016) [5.11]. See Appendix C for a summary.

There were two transition crashes in Phase 1 that Teoh and Kidd chose not to count: a Major PDO crash on February 25, 2015, and an injury crash on August 20, 2015. Teoh and Kidd (2017) [5.1] did not consider it a bias to disregard transition crashes in evaluating AV safety. However, this choice leaves open the possibility for bias. Consider that an AV test driver, with the vehicle in autonomous mode, sees that a crash is imminent and believes it is unpreventable. The test driver could switch the classification of the crash in the CA OL-316 crash report [5.2] from autonomous mode to manual mode by a quick touch of the steering wheel or brake to disengage from autonomous mode (see Appendix C).

Correction Method: Count Transition Crashes as AV Crashes. Downward bias in the AV crash IRR estimate from censoring transition crashes was removed by adding transition crashes to the "police-reportable" autonomous mode crashes in the Teoh and Kidd (2017) [5.1] original data in Table 5.1 (note b). The two crashes in transition (February 25, 2015, and August 20, 2015) during Phase 1 were now counted as crashes in autonomous mode. The number of AV crashes judged as "police-reportable" in Phase 1 (Table 5.1, note b) then became five, not three. The number of AV crashes with an actual police record remained zero (see Chapter 2, Section 2.4.1, Table 2.5).

There is a debatable uncertainty among driving safety experts about (1) whether or not to count transition crashes as autonomous mode crashes and (2) what constitutes a "police-reportable" crash. A "police-reportable" crash can differ from the police department criteria for creating a crash record in whatever city the crash occurred. One way to remove both these uncertainties simultaneously is to estimate the IRRs for all possible AV "police-reportable" crash counts. In other words, estimate the AV crash IRR and its confidence limits for all possible "police-reportable" AV crash counts from zero to five, or seven if transition crashes are counted. However, there were no actual police-recorded crashes (Section 5.1.2.1), so this suggested additional analysis was not further pursued.

5.1.2.3. AV MVMT Site Selection Bias 1: Inclusion of Non-CA Mileage (Table 5.1, Note c)

Teoh and Kidd (2017) [5.1] estimated the total AV mileage in Mountain View from 2009 to 2015 as 1.372111 MVMT (Table 5.1, note c). They obtained this mileage from the December 2015 Google report [5.3], but this mileage also includes mileage from Austin, TX, from July to December 2015. In its July 2015 report [5.3], Google stated, "23 Lexus RX450h SUVs currently self-driving on public streets in Mountain View, CA, and Austin, TX." In its August 2015 report, Google later stated, "We've now been driving in Austin for about two months," meaning July and August 2015. Hence the mileage denominator in the original Teoh and Kidd (2017) [5.1] AV crash rate estimate of 2.19 for Mountain View and vicinity (Table 5.1, note d) is biased high because of the added AV mileage from Austin, TX. There was no government requirement for reporting AV crashes and mileage in Austin, TX, as there is in CA. So although Google did report some crashes in Austin, TX, it may not have reported all AV crashes in its monthly Google reports [5.3]. The inclusion of the non-CA AV mileage means that the Teoh and Kidd (2017) [5.1] AV crash rate estimate of 2.19 in Table 5.1 may have been biased low from its true value.

Correction Method: Remove Non-CA Mileage. The Google CA DMV disengagement report for December 2015 (Google, 2015) [5.12] gives the mileage in CA from September 24, 2014, through November 30, 2015. The Google (2016) [5.12] disengagement report provides the total AV mileage for all states through December 2015. Thus the AV mileage for Austin, TX (and any non-CA AV mileage, if it exists) can be removed from the original AV mileage estimate of Teoh and Kidd (2017) [5.1] in Table 5.1 (note c). The revised Google AV mileage solely in Mountain View and vicinity during Phase 1 is 1.091 MVMT.

5.1.2.4. AV MVMT Site Selection Bias 2: Unspecified AV Mileage in Other Cities (Table 5.1, Note c)

Google's geo-fenced ODD was not just in Mountain View but included cities in Mountain View's immediate vicinity. For example, one of the Major PDO AV crashes occurred in March 2013 in San Jose, CA, and a Minor PDO crash occurred in March 2014 near Belmont, CA. Hence, miles driven in CA locations other than in Mountain View were included in Google's early AV reports [5.3], but only as part of the total AV mileage, not broken out by the city. Accordingly, the AV-exposed crash and mileage counts were not solely from Mountain View. However, the Teoh and Kidd (2017) [5.1] CV crash and mileage counts were all from Mountain View, so there was an additional site selection bias in the AV crash IRR estimate of 2.19 (Table 5.1, note d), based on the original Teoh and Kidd (2017) [5.1] data.

Correction Method: Not Possible. Google has not publicly revealed how many Google AV miles were driven in each CA city in its ODD. (The bulk of the crashes were in Mountain View, but some crashes were in nearby towns or cities too.) Hence, it is not possible to adjust Google AV mileage at the city level. However, the AV mileage in each town or city is known internally in Google because every moment of AV operation from key-on to key-off is GPS-coded. Therefore, it is not possible to correct this city mileage bias with current publicly available information. This second site selection bias remains unadjustable even after correcting all other known biases and errors in Teoh and Kidd's (2017) [5.1] original data in Table 5.1.

However, because most of the AV crashes took place in Mountain View, it is plausible that most of the AV miles occurred there. In addition, AV crash rates may not have differed substantially between the cities near Mountain View and Mountain View itself. Therefore, this second site selection bias from the unknown mileage in nearby cities may not have as extreme consequences as the other biases and errors identified herein for Teoh and Kidd's (2017) [5.1] data in Table 5.1.

5.1.2.5. CV Crash Selection Bias 1: Multi-Vehicle CV Crashes (Table 5.1, Note f)

Teoh and Kidd's (2017) [5.1] method to select CV crashes contained a substantial upward multi-vehicle crash selection bias that substantially inflated their Mountain View "vehicle involvement" count of 9,198 (Table 5.1, note f). Appendix I replicates this bias. As a result, the CV vehicle-involvement rate estimate for Mountain View (Table 5.1, note h) was biased high, which biased the Table 5.1 (note i) Google AV IRR estimate of 0.36 too low.

Correction Method: Count Multi-Vehicle CV Crashes as One Crash. The correction method counts a multi-vehicle CV crash as a single crash rather than two, three, or four crash-involved vehicles. Counting a single crash was the criterion used by Teoh and Kidd (2017) [5.1] for AV crashes when one or more CVs were also involved in the crash. So using the same counting method for CV crashes corrects the multi-vehicle CV crash selection bias in Teoh and Kidd's (2017) [5.1] analysis methods.

5.1.2.6. CV Crash Selection Bias 2: Highway CV Crashes (Table 5.1, Note f)

Teoh and Kidd (2017) [5.1] had a second CV crash selection bias that also upwardly biased their CV police-recorded Mountain View "vehicle involvement" count of 9,198 (Table 5.1, note f). This upward CV crash selection bias was in addition to the upward crash selection bias described in Section 5.1.2.5. Teoh and Kidd (2017) [5.1] counted not only the number of vehicles involved in crashes on city streets of Mountain View but also the number of vehicles involved in crashes on state roads, highways, and interstates within Mountain View city limits. In short, the numerator of Teoh and Kidd's (2017) [5.1] CV crash rate (Table 5.1, note f) included <u>all</u> vehicles involved in all CV crashes on <u>all</u> roads within Mountain View city limits.

This bias occurred because the CV mileage estimates by Teoh and Kidd' (2017) [5.1] did not match how they selected the crashes. Teoh and Kidd's (2017) [5.1] CV mileage estimate (Table 5.1, note g) included mileage <u>only</u> on Mountain View city streets. Their CV mileage estimate did <u>not</u> include the mileage on state roads, highways, and interstates within Mountain View city limits (see Section 5.1.2.10). The result was a mismatch in the looser criterion for CV "crash involvement" in the numerator and the stricter criterion for CV mileage in the denominator of the CV crash rate estimate. Consequently, there was an upward bias in the CV "crash involvement" rate estimate. There were extra highway crashes in the numerator that were not matched by the highway mileage in the denominator. This upward bias in the CV crash rate caused a downward bias in the AV crash IRR, adding additional downward bias to the AV crash IRR in addition to the multi-vehicle downward IRR bias described in Section 5.1.2.5.

Thus, there was a substantial mismatch in the criteria that Teoh and Kidd (2017) [5.1] used for the numerator and denominator in their CV "crash involvement" rate estimate (Table 5.1, note h).

The numerator of Teoh and Kidd's CV crash-involved rate estimate. Passenger vehicles involved in crashes (Table 5.1, note f) were counted on all roads within the Mountain View city limits. Specifically, Teoh and Kidd's count of 9,198 crash-involved CVs was on all roads (Table 5.1, note f) within Mountain View city limits. These roads included not just city streets but also state highways and interstates within the Mountain View city limits. That is, they counted the vehicles involved in crashes on all roads within the city limits.

The denominator of Teoh and Kidd's CV crash-involved rate estimate. But the mileage in the denominator of their CV crash rate estimate was based on Mountain View city streets alone. Indeed, mileage only on city streets is the only available mileage in the HPMS reports [5.4] for Mountain View, which is what Teoh and Kidd (2017) [5.1] cited as the source for their Mountain View mileage estimate of 1,517.93 MVMT (Table 5.1, note g).

In sum, Teoh and Kidd wrongly selected crash-involved vehicles on roads that included crashes on state highways and interstates within Mountain View city limits. That is, their vehicle-involvement CV crash count of 9,198 (Table 5.1, note f) in the numerator of the CR crash rate included crashes on all roads within Mountain View. But their CV mileage estimates in the denominator were for city streets only.

Using different road types for crashes and mileage is another form of site selection bias. This bias upwardly biased Teoh and Kidd's (2017) [5.1] CV crash involvement rate estimate of 6.06 for Mountain View (Table 5.1, note h). This upward bias in the CV crash involvement rate, in turn, downwardly biased the AV IRR estimate of 0.36, based on Teoh and Kidd's (2017) [5.1] original data in Table 5.1.

Correction Method: Remove CV Crashes on Highways. This second crash selection bias in Teoh and Kidd's (2017) [5.1] CV crash rate estimate (Table 5.1, note h) was removed. The selection criterion for the numerator of the CV crash rate estimate (Table 5.1, note f) was revised to match the selection criterion for the denominator of the CV crash rate estimate (Table 5.1, note g). Note that the original CV mileage estimate of 1517.93 MVMT was restricted to Mountain View city streets. Therefore, the revised count of passenger "vehicle involvement" crashes (Table 5.1, note f) was also restricted to Mountain View city streets. Crashes on state highways and interstates with portions within Mountain View city limits were excluded, thereby removing this second upward bias in the CV crash rate and avoiding an upward bias in the CV-involved crash count.

Two methods can query the SWITRS database [5.5] to restrict the CV-involved crashes solely to city streets:

1. In an online query, deselect the checkbox "INCLUDE COLLISIONS THAT OCCURRED ON STATE HIGHWAY (interstate, U.S., state route)." This checkbox refers to "collisions which occur on or are associated with a state highway."
2. In a submitted query, request the raw data files using a checkbox on the SWITRS [5.5] Web page. The raw data files for the period of interest will then be sent to the submitter via email. Then the crashes on state and interstate highways within Mountain View city limits can be filtered out by hand using Excel or a database program.

Removing the second CV crash selection bias by one of these two methods has the added benefit of a better ODD match of the Google AV crash and mileage estimates. Google AVs are geo-fenced to a particular ODD while in autonomous mode. The reason is that all the Google AV crashes except one occurred on city streets. The exception was an early Major PDO sideswipe crash at a 63 mph speed on a highway in San Jose in March 2013

(Google Reports, March 2013 [5.3] and Blanco et al., 2016, Table 12 [5.10]). It is conjectured that Google geo-fenced its AVs after that high-speed highway crash to drive only on slower-speed city streets when in autonomous mode. There was no other high-speed highway crash after that initial crash. Therefore the Google AV crashes were almost all on city streets under the jurisdiction of the Mountain View police department (including El Camino Real). The added benefit of restricting CV crashes to city streets also provides a better match to the Google AV ODD.

Caveat: In filtering out state and interstate highway crashes and mileage, an exception was made for certain roadways that were state highways by historical precedents, such as CA State Route 82, but which is now effectively a city street known as "El Camino Real" in Mountain View with a 35 mph speed limit and intersections, crosswalks, etc. like any other city street. El Camino Real is also now under the Mountain View city police jurisdiction and not the CHP. El Camino Real (CA State Route 82) is clearly within the Google AV ODD and planned test route within that ODD because many Google AV crashes occur there [5.2, 5.3].

5.1.2.7. CV Crash Selection Bias 3: Counting Passenger Vehicles Only (Table 5.1, Note f)

Teoh and Kidd's (2017) [5.1] original CV-involved crash count of 9,198 (Table 5.1, note f) was only for passenger vehicles involved in crashes. However, their original mileage estimate of 1,517.937 MVMT (Table 5.1, note g) was for all vehicle types. Indeed, mileage from all vehicle types is the only mileage available in the HPMS database [5.4]. The result of mismatching the vehicle and mileage criteria was a downward bias in Teoh and Kidd's (2017) [5.1] original CV "crash involvement" rate estimate of 6.06 (Table 5.1, note h). That bias, in turn, caused an upward bias in the AV crash IRR estimate of 0.36 (Table 5.1, note f). Teoh and Kidd (2017, p. 58) [5.1] acknowledged this bias in their article by claiming that it was okay to ignore this bias because "Passenger vehicles comprise the vast majority of VMT."

Correction Method: Count All CV Types. The crash selection bias from tabulating only passenger vehicles rather than tabulating all types of vehicles involved in crashes to obtain the CV-involved crash count of 9,198 (Table 5.1, note f) was eliminated. This correction was accomplished by tabulating all vehicle types for the CV crash count, rather than just passenger vehicles.

Note that all crashes of all CV types (not just passenger vehicles) were tabulated in the five other critically reviewed studies in this book (see Chapters 2, 3, 4, 6, and 7). It would also be preferable to tabulate crashes of only passenger vehicles in the Mountain View CV column (Table 5.1, note 3) to match Google AVs, which were all passenger vehicles. However, there are no mileage estimates for individual vehicle types in the HPMS database [5.4]. Therefore, there is no valid alternative but to include all vehicle types in the final CV crash count estimate.

In summary of the CV crash count, the original Mountain View CV passenger vehicle crash-involvement count by Teoh and Kidd (2017) [5.1] was 9,198 (Table 5.1, note f). After eliminating the three CV crash selection biases documented in Sections 5.1.2.5, 5.1.2.6, and 5.1.2.7, the revised Mountain View CV crash count is 1,993, a substantial decline from the original Teoh and Kidd (2017) [5.1] CV crash count estimate. Correcting this bias will lower the CV crash rate estimate from 6.06 (Table 5.1, note h), which will upwardly adjust the original AV crash IRR estimate of 0.36 (Table 5.1, note i).

5.1.2.8. CV MVMT Minor Update for 2015 (Table 5.1, Note g)

Teoh and Kidd (2017, p. 58) [5.1] repeated the 2014 MVMT (209.256) in Mountain View, CA (HPMS, 2014) [5.4] for 2015 because they said, "At the time of this study, California

VMT data were not available for 2015, so the 2014 figures were used as estimates in calculating crash rates for 2010–2015 in aggregate." Thus, Teoh and Kidd's (2017) [5.1] original total mileage estimate for 2010–2015 in Mountain View was 1,517.93 MVMT (Table 5.1, note g).

Correction Method: Use Actual 2015 Data. The actual 2015 MVMT for Mountain View was later published (HPMS, 2015). That figure converted to 224.268 MVMT (HPMS, 2015) [5.4], or a 15.012 MVMT increase (7.2%) from Teoh and Kidd's (2017, p. 58) [5.1] repetition of the 2014 MVMT estimate of 209.256 for their 2015 MVMT estimate. The corrected total MVMT estimate for 2010–2015 in Mountain View was therefore 1,532.95 MVMT, only about a 1% increase from Teoh and Kidd's (2017) [5.1] original total MVMT estimate for 2010–2015 in Mountain View of 1,517.93 MVMT (Table 5.1, note g).

Nonetheless, this increase in the total CV mileage biased the Teoh and Kidd (2017) CV crash rate estimate of 6.06 (Table 5.1, note h) slightly low, which would bias the AV crash IRR estimate slightly high. Therefore, the current reanalysis used the actual 2015 mileage. Accordingly, the total CV mileage estimate for 2009–2015 for Mountain View was updated to 1,532.95 MVMT, a 1% increase.

5.1.2.9. CV MVMT Site Selection Bias (Table 5.1, Note g)

Table 5.1 compared the AV crash rate to the CV "crash involvement" rate, using the original data in Teoh and Kidd (2017) [5.1]. The Google AV ODD was not solely within Mountain View, CA, because some AV crashes were in adjoining cities to Mountain View (e.g., Palo Alto). In the early Google tests [5.3], one crash even occurred on Highway 680S in San Jose (south of Mountain View) at 63 mph (Blanco et al., 2016, p. 59) [5.10]. Another Google AV crash occurred on Highway 101N near Belmont, CA (north of Mountain View) (Blanco et al., 2016, p. 59) [5.10]. However, Teoh and Kidd's (2017) [5.1] CV crash involvement rate estimate (Table 5.1, note g) was based solely on Mountain View data.

Correction Method: None Available. The CV crash rate estimate should be expanded to include all the cities in the Google AV ODD. However, the Google AV crash rate cannot be estimated for specific cities outside Mountain View because Google did not specify the AV mileage in each city in its ODD. Therefore, it is impossible with current publicly available data to adjust for the CV MVMT site bias. The original Teoh and Kidd (2017) [5.1] analysis and the present reanalysis have a remaining CV mileage site bias that cannot be adjusted without new information from Google.

In short, the comparison of Google AV crash rates to CV crash rates in Mountain View (or to Santa Clara County, the State of California, or the USA nationally) contains a non-adjustable site bias. This site bias will remain until and unless Google/Waymo and other companies publicly release the specific AV mileage in every city in their AV ODDs.

It is also not determinable from the information supplied by Google and other AV companies whether the test area for their AVs even corresponded to their ODDs. Although the Google AVs are geo-fenced to operate in autonomous mode solely within an ODD, AVs likely traversed only a portion of their ODD on any given day because of the particular test requirements for that day. In sum, although Google correctly maintains that their AV mileage is mileage solely in autonomous mode, Google does not specify what AV mileage is included on what specific road types within its ODD or even within what cities.

In sum, correcting Teoh and Kidd's (2017) [5.1] CV MVMT site bias cannot be done with publicly available information. Correction of this bias requires additional information to be released by Google and other AV companies of the AV mileage in each city in their AV ODD.

5.1.2.10. CV MVMT City Street Site Selection Bias (Table 5.1, Note g)

(Note: This bias is closely related to the Highway CV crash bias described in Section 5.1.2.6.)

Teoh and Kidd (2017, Table 1) [5.1] give the source for their Mountain View CV mileage estimate for 2009–2015 of 1,517.937 MVMT (Table 5.1, note g) as the HPMS (2009–2014) reports [5.4]. However, the HPMS mileage for cities that are not also a county (such as San Francisco) is miles driven on city streets only. The mileage does not include miles driven on state and interstate highways within city limits. This restriction to city street mileage includes Mountain View and the few other cities in the ODD of the Google AVs in Santa Clara County, CA. (The only condition the HPMS database gives the miles driven on all roads within a city is when the city is the whole county, such as San Francisco).

In other words, the HPMS mileage [5.4] for any CA city (such as Mountain View, which is not also a county) is the mileage on city streets only. The mileage does not include mileage on state highways and Federal interstates within the city limits.

Hence, there was a substantial mismatch between the selection criterion that Teoh and Kidd (2017) [5.1] used for (1) CV-involvement crashes (not just city streets but also state and interstate highways within Mountain View) and (2) CV mileage (city streets only). This mismatch in the selection criterion for CV crash locations and CV mileage upwardly biased Teoh and Kidd's (2017) [5.1] CV-involved crash rate estimate of 6.06 (Table 5.1, note h), and downwardly biased the AV crash IRR estimate of 0.36 (Table 5.1, note j) based on Teoh and Kidd's (2017) [5.1] original data.

Correction Method: None Needed After Removal of Highway CV Crashes. There are two potential methods to remove the city street site selection bias and make the CV crash and mileage venues the same.

The first method to remove the bias is to add the state highway and freeway mileage within Mountain View city limits to the CV mileage estimate (Table 5.1, note g). Unfortunately, this first method is impossible because there is no known way to determine the highway and freeway mileage within Mountain View city limits with publicly available mileage databases. For example, the number of miles driven on highways and freeways within the geographic boundaries of Mountain View (or any other CA city that is not a county) is not available in the HPMS database [5.4], nor in any other known public database. A mileage estimate that includes highway and freeway miles within a geographic boundary exists in the CA public databases only for counties, not cities. (As mentioned previously, the only exception is for cities that form an entire county, such as San Francisco).

A second method that indirectly removes the CV MVMT site selection bias is accomplished with the correction method in Section 5.1.2.6. That correction method removes highway CV crashes, restricting the CV crash count (Table 5.1, note f) to city streets alone. CV crashes on city streets alone then matches the CV MVMT denominator (Table 5.1, note g), which is already city streets alone. This second method accomplishes that by removing the state and interstate highway crashes from the CV crash rate estimate (Table 5.1, note f). This correction method is the same as the correction method in Section 5.1.2.6. In other words, after applying the correction method in Section 5.1.2.6, both the CV crashes and mileage were on city streets, so they matched. Therefore, the Mountain View CV MVMT site selection bias was eliminated without adjusting the mileage further. CV crashes and mileage were already both from city streets alone after correcting the CV crash selection bias in Section 5.1.2.6. Hence, no further change in the CV MVMT was necessary to remove the CV MVMT city street site selection bias.

5.1.3. Teoh and Kidd's (2017) Partly-Corrected Phase 1 Data and AV IRR

After correcting the Teoh and Kidd (2017) [5.1] data for its known biases and errors identified in Section 5.1.2, a more valid estimate of the AV IRR of 2.11 is given by the partly-corrected data in Table 5.2.

TABLE 5.2 Teoh and Kidd's (2017) Phase 1 data: Partly-corrected.

	Google AV[a]	Mountain View CV[e]
Crashes	3[b]	**1,993**[f]
MVMT	**1.091438**[c]	**1,532.95**[g]
Crash rate	2.75[d]	1.30[h]
IRR (95% CI)[i]	2.11 (0.54–5.8), $P = 0.23$	

The crash rates of Google AVs compared to 2009–2015 Mountain View CVs.

A **bold** number in the 2×2 matrix above means no identified error or bias.

Notes:

[a] Google AV data on city streets in Mountain View, CA, and vicinity from 2009 to December 31, 2015 (excludes Austin, TX).

[b] Same as Table 5.1.

[c] Google AV mileage on city streets in Mountain View and vicinity. (Austin, TX mileage excluded. See Section 5.1.2.3).

[d] Google AV city street crash rate (Mountain View and vicinity).

[e] Same as Table 5.1.

[f] Total police-recorded CV crashes on Mountain View city streets (SWITRS, 2009–2015) [5.5]. (Additional vehicles involved in a CV crash were excluded; see Section 5.1.2.5. Also, crashes on state roads, highways, and freeways were excluded; see Section 5.1.2.6.

[g] CV mileage on city streets in Mountain View, 2009–2015 (HPMS, 2009–2015) [5.4]. (The 2015 CV mileage was updated to the actual mileage instead of repeating the 2014 mileage per Section 5.1.2.8.)

[h] CV crash rate on Mountain View city streets.

[i] Same as Table 5.1.

The Google AV mileage (Table 5.2, note c) decreased after censoring non-CA mileage (Section 5.1.2.3). Therefore, the Google AV crash rate increased from 2.19 (Table 5.1, note d) to 2.75 (Table 5.2, note d).

The Mountain View CV events decreased from 9,198 (Table 5.1, note f) to 1,993 (Table 5.2, note f) after removal of the bias from counting every passenger vehicle in a multi-vehicle crash (Section 5.1.2.5).

The partly-corrected data in Table 5.2 estimate an AV crash IRR of 2.11 (Table 5.2, note i) compared with the IRR estimate of 0.36 in Table 5.1 (note i) based on the original Teoh and Kidd (2017) [5.1] data. This upwardly revised IRR estimate of 2.11 indicates an increase of about 111% for the Google AV "police-reportable" crash rate in Mountain View and vicinity versus the police-recorded CV crash rate in Mountain View, CA. Nonetheless, the Table 5.2 IRR 95% confidence range (0.54–5.8) would suggest that a "police-reportable" crash rate ranging from a 46% decrease to a 475% increase for Google AVs versus passenger CVs is also reasonably compatible with the partly-corrected Phase 1 data of Teoh and Kidd (2017) [5.1] in Table 5.2.

Why Only Partly-Corrected?: AV Crash Uncertainty Remains

The reason Table 5.2 is only partly corrected is there is a remaining uncertainty in the three AV crashes (Table 5.2, note b) that Teoh and Kidd (2017) [5.1] judged to be "police-reportable" (Table 5.1, note b). As noted in Section 5.1.2.1, the actual number of "police-reportable" AV crashes could be anywhere from 0 to 5. For example, it might be five if Teoh and Kidd (2017) [5.1] had included transition crashes in their definition of AV crashes (see Section 5.1.2.2 and Appendix C). Or, the AV crashes could be zero if Teoh and Kidd (2017) [5.1] had not relied on their judgment as to what "police-reportable" is but had checked the SWITRS database [5.5] to find out how many of the AV crashes had an actual police record (Section 2.4.1 and Table 2.5).

Successively inserting the possible values from 0 to 5 of "police-reportable" AV crashes in Table 5.2 (note b), the possible full range of AV crash IRR estimates for Teoh and Kidd's (2017) [5.1] Phase 1 ranges from 0.0 (for AV crashes = 0) to 3.49 (for AV crashes = 5). The crash IRR range (from 0 to 5 "police-reportable" AV crashes) is thus from a 100% AV rate decrease (for zero AV crashes) to a 249% AV rate increase (for five AV crashes) compared to the Mountain View CV crash rate. Nonetheless, these corrected results would suggest that, based on the maximum possible range of all five of the 95% confidence limits, the "police-reportable" AV crash rate ranges from a 100% decrease to a 716% increase for Google AVs versus human-driven CVs during Teoh and Kidd's (2017) [5.1] Phase 1. This wide range of possible IRR estimates, from decreased risk to increased risk, is therefore compatible with the corrected Teoh and Kidd (2017) [5.1] Phase 1 data in Table 5.2.

However, in Chapter 2 (Section 2.4.1 and Table 2.5), it was shown that none of the Google AV crashes had a police record, not just the three predicted by Teoh and Kidd (2017) [5.1]. Therefore, the method of assessing all possible values of "police-reportable" crashes in the previous paragraph was not pursued further because none of the AV crashes had an actual police record. (Note it could still be possible that these crashes had a PAR filled out by the police station, but the local department never uploaded a crash record for it to the state police database.)

5.1.4. Discussion and Conclusion to Teoh and Kidd's (2017) Phase 1

Ten biases and errors in Teoh and Kidd's (2017) [5.1] data analysis were described in the Sections indicated:

1. Two AV crash selection biases (Sections 5.1.2.1 and 5.1.2.2).
2. Two AV MVMT site selection biases (Sections 5.1.2.3 and 5.1.2.4).
3. Three CV crash selection biases (Sections 5.1.2.5, 5.1.2.6, and 5.1.2.7).
4. One CV MVMT update (Section 5.1.2.8).
5. Two CV MVMT biases (Sections 5.1.2.9 and 5.1.2.10).

Adjustments and corrections to the AV MVMT increased the "police-recordable" AV crash rate estimate from 2.19 (Table 5.1, note d) to 2.75 (Table 5.2, note d) "police-recordable" AV crashes per MVMT. Adjustments and corrections to the CV crashes and MVMT substantially reduced the CV crash rate estimate for Mountain View from 6.06 (Table 5.1, note h) to 1.30 (Table 5.2, note h) police-recorded CV crashes per MVMT. After these adjustments and corrections, the estimated AV total crash IRR substantially increased from 0.36 in Table 5.1 (note j) to 2.11 in Table 5.2 (note j).

A limitation of the Phase 1 analysis in Section 5.1 is a remaining uncertainty in Teoh and Kidd's (2017) [5.1] definition and judgment of a "police-reportable" AV crash.

One method to reduce this uncertainty is to estimate the AV IRR for all possible "police-reportable" AV crashes from zero to five. This method was applied to the Phase 1 data of Teoh and Kidd (2017) [5.1] near the end of Section 5.1.3. Even so, it was still not possible to determine with a reasonable degree of certainty as to whether AVs increased or decreased crash rates relative to human-driven CVs with the Phase 1 data with the methods used by Teoh and Kidd (2017) [5.1]. With zero actual "police-reported" AV crashes, the AV crash rate is lower than the CV crash rate. However, there is considerable uncertainty as to why so few AV crashes have police records (see Appendix E).

For such reasons, the current review concludes that the corrected Phase 1 data of Teoh and Kidd (2017) [5.1] in Table 5.2 fail to support their conclusion that "Google self-driving cars…are safer than conventional human-driven passenger vehicles." AVs may be safer than CVs for Severe crashes but not Major, or Minor PDO crashes, as suggested by the Chapter 3 analysis. But the data in "Phase 1" and the analysis methods used by Teoh and Kidd (2017) [5.1] do not allow for that conclusion.

5.2. Phase 2 "All States": January 1, 2016, to August 31, 2016

5.2.1. Summary of Teoh and Kidd's (2017) Phase 2

Teoh and Kidd (2017) [5.1] designated the first eight months of 2016 as their Phase 2 "All States" analysis period. During all or part of that period, AV tests took place in the vicinity of four cities in four states: Mountain View, CA; Austin, TX; Chandler, Arizona; and Kirkland, WA. However, Teoh and Kidd (2017) [5.1] did not estimate the CV miles driven in the three non-CA cities in Phase 2, and so a CV crash rate could not be estimated for Phase 2 as it was for Phase 1 (Table 5.2, note f).

However, the AV crash rate could be estimated in Phase 2 from the AV miles and crashes data provided by Teoh and Kidd (2017, Table 2) [5.1]. The Google test fleet drove 596,967 miles, with nine autonomous-mode crashes reported in Google's 2016 monthly reports [5.3]. Furthermore, Teoh and Kidd (2017) [5.1] judged six of those AV crashes in Phase 2 to be "police-reportable." As a result, Teoh and Kidd (2017, p. 59) [5.1] state, "The Google police-reportable crash rate during the first eight months of 2016 was 10.05 per million VMT."

Teoh and Kidd (2017) [5.1] did not mention that the AV crash rate during Phase 2 (10.05 AV crashes per MVMT) was more than four times larger than during Phase 1 (2.19 AV crashes per MVMT) (Table 5.1, left data column). However, Teoh and Kidd (2017) [5.1] did note that after they combined the "police-reportable" crashes and mileage in Phases 1 and 2, the crash rate for their "All States" from 2009 to August 31, 2016, was 4.57 AV crashes per MVMT (Teoh and Kidd, 2017, Table 2) [5.1]. Thus, this Phase 2 AV crash rate estimate was approximately double their Phase 1 "Mountain View" AV crash rate of 2.19 (Table 5.1, note f).

This doubling of the AV crash rate is difficult to reconcile with Teoh and Kidd's (2017) [5.1] conclusion that "Google self-driving cars…are safer than conventional human-driven passenger vehicles." Teoh and Kidd (2017, p. 60) [5.1] attempted to reconcile this contradiction with a hypothesis to explain why the Google AV crash rate approximately doubled for Phase 2 compared to Phase 1. They state that "The doubling of the likelihood that Google's crash incidents were deemed police-reportable, attributable to the first eight months of

2016, may reflect increased usage on higher speed roads other than interstates/freeways or more aggressive testing practices."

Teoh and Kidd (2017) [5.1] also noticed another difference between Phases 1 and 2 in the proportion of crashes they judged "police-reportable." They state (p. 59), "Before 2016, there were ten crash incidents in autonomous mode, three of which were judged police-reportable, versus 6 of 9 during January to August 2016." Thus, they noted an approximate doubling of the proportion of crashes judged "police-reportable" from Phase 1 to Phase 2. That is, not only did the number of crashes double in Phase 2, the severity of those crashes also increased. However, again, Teoh and Kidd (2017) [5.1] did not discuss the implications of this doubling of the "police-reportable" crash proportion for Phase 2 on their conclusion that Google AVs are safer than CVs.

5.2.2. Critical Analysis of Teoh and Kidd's (2017) Phase 2

Teoh and Kidd's (2017) [5.1] Phase 2 analysis avoided many of the biases in their Phase 1 analysis (Section 5.1.2). In Phase 2, they did not attempt to estimate a CV-involved crash rate, presumably because the Google reports [5.3] did not supply mileage data for their non-CA test sites. However, Teoh and Kidd (2017) [5.1] did directly compare the AV crash rates in Phase 2 to Phase 1 (Section 5.2.1). Table 5.3 puts Teoh and Kidd's (2017) [5.1] Phase 2 versus Phase 1 data into an epidemiologic design for a rigorous comparison (see Chapter 1, Sections 1.1 and 1.2).

Table 5.3 compares Teoh and Kidd's (2017) [5.1] AV data for Phase 2 with their uncorrected Phase 1 AV data (Table 5.1, left data column). The "All States" Phase 2 "police-reportable" AV crash rate estimate for the Google AV test was 10.05 crashes per MVMT (Table 5.3, note d). On the other hand, the Mountain View Phase 1 original Teoh and Kidd (2017) [5.1] "police-reportable" AV crash rate estimate was 2.19 crashes per MVMT (Table 5.3, note h; also Table 5.1, note d). Hence Table 5.3 shows that the "police-reportable" AV crash rate estimate increased from 2.19 per MVMT in Phase 1 (note h) to 10.05 per MVMT in Phase 2 (note d), a 4.60 multiple, or 360% increase. The bottom row of Table 5.3 compares the

TABLE 5.3 Epidemiologic analysis of Teoh and Kidd (2017) Phase 2 versus Phase 1 AV crash rates.

	Phase 2 AV[a]	Phase 1 AV[e]
Crashes	**6**[b]	**3**[f]
MVMT	0.596967[c]	1.372111[g]
Crash rate	10.05[d]	2.19[h]
Phase 2/1 IRR[i]	4.60 (95% CI 1.14-22.5), P = 0.0312	

A **bold** number in the 2×2 table above means no identified error or bias.

Notes:

[a] Phase 2 Google AV driving data in four states (January 1, 2016, to August 31, 2016). Calculated as the difference between "All States" and "Mountain View" crashes and mileage in Table 2 in Teoh and Kidd (2017) [5.1].

[b] AV crashes "judged to be police-reportable" in four states.

[c] AV MVMT in four states.

[d] AV crash rate in four states.

[e-h] Same as Table 5.1 (notes a-d).

[i] Same as Table 5.1.

Phase 2 to Phase 1 "police-reportable" AV crash data using the IRR design in Chapter 1 (Table 1.1), which estimates an IRR of 4.60.

Teoh and Kidd (2017) [5.1] correctly recognized that it was not just the AV crash rate that substantially increased from Phase 1 to Phase 2; the severity of the AV crashes also increased. They note that the percentage of AV crashes that were serious enough to be judged "police-reportable" doubled from 30% in Phase 1 (3 of 10 crashes) to 67% in Phase 2 (6 of 9 crashes). These sample sizes are too small for a valid test using traditional statistics. With epidemiologic methods, small sample sizes still made a valid estimate of the substantial crash IRR effect size of 4.60 (Table 5.3, note i) comparing Phase 2 versus Phase 1 AV crash rates. This 360% increase in the AV crash rate and crash severity in Phase 2 versus Phase 1 appears inconsistent with Teoh and Kidd's (2017) [5.1] conclusion that "Google self-driving cars...are safer than conventional human-driven passenger vehicles." Figure 2.2A in Chapter 2 plots the Schoettle and Sivak (2015) [5.8] AV crash rate from February to September 2015. The upward trend in the AV crash rate reflects a similar upward trend in the CV crash rate over the same period (Figure 2.2B). (It is also inversely correlated with the downward trend in the AV test driver cumulative disengagement rate in Figure 2.2C).

In other words, the increased AV crash rate for Teoh and Kidd's (2017) [5.1] data in their Phase 2 test period (the first eight months of 2016) compared to the Phase 1 period (from 2015 and before) is consistent with the upward trend in the Google AV crash rate in 2015 (Figures 2.1 and 2.2B in Chapter 2). The 2015 upward trend in the AV cumulative crash rate was evident in Figures 2.1 and 2.2A. Therefore, the Google AV crash rate increase in 2016 just continued the upward trend in the AV cumulative crash rate trend that began in 2015. This hypothesis is examined more fully in Section 5.3 below.

5.3. Trends in Teoh and Kidd's (2017) Data

There can be additional viewpoints and corrections of the data and analysis methods used by Teoh and Kidd (2017) [5.1]. However, a simple method summarizes the Google AV data examined by Teoh and Kidd (2017) [5.1] in a single graph, as per Figure 5.1 (next page).

Figure 5.1 compares the Google AV cumulative crash rate to itself, as did Figure 2.1. Hence, the trends are not subject to crash selection bias or site selection bias when AV crash rates are compared to a CV crash rate.

By visual inspection of Figure 5.1, the "police-reportable" Google AV cumulative crash rate was flat between the March 2013 and July 1, 2015 crash. Then an upward trend in the cumulative crash rate began that continued until the crash on August 22, 2016, near the end of the Teoh and Kidd (2017) [5.1] study period on August 31, 2016. This same pattern is even more evident in Figure 2.1 in Chapter 2, where the Google AV Minor PDO and transition crashes were included that were not included in Figure 5.1.

This upward trend in the Google AV cumulative crash rate in 2016 is similar to the upward trend observed in 2015 in the Schoettle and Sivak (2015) [5.8] data, as shown in Figures 2.1 and 2.2 in Chapter 2.

Figure 5.1 confirms, by visual inspection, the observation by Teoh and Kidd (2017) [5.1] that the "police-reportable" AV crash rate increased from Phase 1 in 2009–2015 to Phase 2 in 2016. The current analysis tested this observation in Table 5.3 and found it valid: Phase 2/1 IRR = 4.60 (95% CI 1.14–22.5), $P = 0.031$. What can explain this increase?

FIGURE 5.1 Teoh and Kidd's (2017) [5.1] AV "police-reportable" cumulative crash rate by date of the crash, Phases 1 and 2. AV cumulative crash rate (cumulative AV crashes ÷ cumulative AV MVMT at the crash date) for each of the nine Google AV "police-reportable" crashes tabulated by Teoh and Kidd (2017) [5.1] during their Phases 1 and 2 (2009-8/31/16). **Large dark-green circles:** Severe (Severity Level 1) crashes (7/1/15, 3/14/16). **Medium light-green circle**: Major PDO (Severity Level 2) crashes. Minor PDO crashes (Severity Level 3) and crashes during the transition from autonomous to manual mode (on 2/25/15 and 8/20/15) are not shown because Teoh and Kidd (2017) [5.1] excluded them from their study.

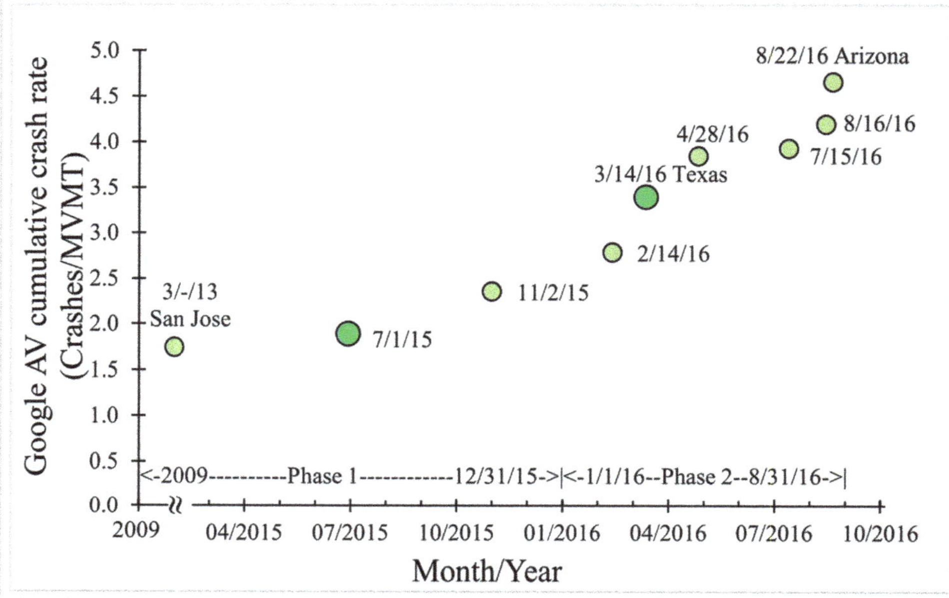

One possible explanation is that the increased Phase 2 "police-reportable" AV crash rates arose from the expansion of the Google AV test sites from Mountain View, CA, and vicinity, to the three states outside CA (AZ, TX, and WA). However, for the six "police-reportable" AV crashes in Phase 2 starting January 1, 2016, the "police-reportable" AV cumulative crash rate also increased for the four AV crashes just in Mountain View, CA on February 14, 2016; April 28, 2016; July 15, 2016; and August 16, 2016. That is, the increase was not just for the crashes outside CA: The TX AV crash on March 14, 2016, and the AZ AV crash on August 22, 2016, as evidenced by visual inspection of Figure 5.1. Therefore, this possible explanation for the increased Phase 2 "police-reportable" AV crash rates was rejected. In addition, the crashes in TX on March 14, 2016, and AZ on August 22, 2016, are excluded in subsequent analyses in this Chapter because Google did not provide AV mileage data for states outside CA.

Teoh and Kidd (2017) [5.1] provided a more viable hypothesis for this increase: "increased usage on higher speed roads other than interstates/freeways or more aggressive testing practices." This hypothesis of Teoh and Kidd (2017) [5.1] to explain the increased "police-reportable" AV crashes in 2016 was also considered in Chapter 2 to explain the rise in 2015 in the Google AV cumulative crash rate (see Hypothesis 1 in Chapter 2, Section 2.4.2.2.1).

Three additional hypotheses to explain the upward trend in 2015 in the cumulative crash rate were advanced in Chapter 2, Section 2.4.2.2. Hypothesis 2 is "negative learning by the Google ADS neural network" and is not repeated here (see Chapter 2, Section 2.4.2.2.2, and Appendix M). Instead, Hypotheses 3 and 4 are investigated in this Chapter to explain what is evident from visual inspection of Figure 5.1: A continuing upward trend in 2016 in the AV cumulative crash rate.

Hypothesis 3 in Section 2.4.2.2.3 was that the 2015 upward trend in the Google AV cumulative crash rate seen in Figures 2.1 and 2.2A was not an actual effect but was a bias arising from the increased probability of a CV crash in Mountain View. This hypothesis was tested using the data in Figure 2.2B in Chapter 2 and found to be plausible. It is tested in this Chapter with the 2016 data because the Google "police-reportable" AV cumulative crash rate continued to trend upward in 2016, as evidenced by visual inspection of Figure 5.1.

Hypothesis 4 in Section 2.4.2.2.4 is that a downward trend in the Google test drivers' cumulative disengagement rate causes an upward trend in the AV cumulative crash rate. Hypothesis 4 was tested in Chapter 2 in Figure 2.2C with the 2015 Google AV data and found plausible. Hypothesis 4 is tested in this Chapter with the 2016 Google "police-reportable" AV crashes. This Chapter also extends Hypothesis 4 to include ADS disengagements as well as test driver disengagements. In this Chapter, the hypothesis is tested that an increase in the ADS cumulative disengagement rate was associated with an increase in the "police-reportable" AV cumulative crash rate.

Hypothesis 3 and 4 are tested using the data in Figure 5.2, which extends the Google AV data into 2016, the range for the Phase 2 Google "police-reportable" AV crashes investigated by Teoh and Kidd (2017) [5.1].

5.3.1. The Trend in AV Cumulative Crash Rate

Figure 5.2A plots the AV cumulative crash rate for the six "police-reportable" Google AV crashes in Mountain View at the dates on the X-axis for 2015–2016, as shown. The linear regression fit ($r = 0.99$, df = 4, $P = 0.00005$) confirms the observed upward trend in the AV cumulative crash rate in Figure 5.2A. It also extends into 2016 a similar finding for the 2015 AV cumulative crash rate in the reanalysis of the data of Schoettle and Sivak (2015) [5.8] (see Chapter 2, Section 2.4.2.2 and Figure 2.2A).

5.3.2. The Trend in CV Cumulative Crash Rate

Figure 5.2B plots the Mountain View CV cumulative crash rate for all police-recorded CV crashes at quarter-end for the same period as Figure 5.2A. The denominators of the CV crash rates were estimated using the cumulative Mountain View city street mileage per month calculated from the monthly CV mileage data in the HPMS database [5.4]. The numerators of the CV crash rates were estimated using the cumulative Mountain View city street CV crash data per month tabulated from the monthly crash data in the SWITRS database [5.5]. Figure 5.2B indicates that the police-recorded CV crash rate has an upward trend from the second quarter of 2015 to the third quarter of 2016 (blue hatched squares). The dotted line is the linear regression fit ($r = -0.98$, df = 12, $P = 6.9E{-10}$). This upward trend confirms and extends into 2016 a similar finding for the 2015 AV crash data of Schoettle and Sivak (2015) [5.8].

5.3.3. The Effect of the CV Cumulative Crash Rate on the AV Cumulative Crash Rate

The CV cumulative crash rate in panel B was interpolated at the Google AV "police-reportable" crash dates in panel A. Then, the linear regression fit between the AV cumulative crash rate in panel A and the interpolated points from panel B gives $r = 0.96$ (df = 4, $P = 0.002$). This fit confirms that the upward trend in the Google AV cumulative crash rate is consistent with an upward trend in the CV cumulative crash rate in Mountain View in 2015–16. That is, the upward trend in the Google AV cumulative crash rate in Figure 5.2A may have been caused by a general increase in the probability of a crash in this period and location, as evidenced by the upward trend in the CV cumulative crash rate in Figure 5.2B.

The influence could not have been in the reverse direction from A to B (an increase in "police-reportable" AV crashes causing an increase in the police-recorded crash count) because there were very few "police-reportable" AV crashes compared to CV crashes. Moreover, there were even fewer (if any) actual police-recorded AV crashes (see Section 2.4.1).

In sum, this effect of the CV cumulative crash rate on the AV cumulative crash rate confirms the 2015 result (Chapter 2, Section 2.4.2.2.3, Figures 2.2A and 2.2B) and extends it into 2016.

This additional confirmation of Hypothesis 3 with the 2016 data makes it less likely that Hypothesis 2 in Section 2.4.2.2.2 (negative learning by the Google neural network) is needed to explain why the Google AV cumulative crash rate rose throughout 2015 and 2016. Instead, a more straightforward explanation is that the Google AV fleet simply exhibited an upward trend in its cumulative crash rate during this period because of a general increase in the probability of a crash in Mountain View. This overall rise in crash probability was evidenced by the upward trend in the CV cumulative crash rate in Mountain View.

5.3.4. The Trend in the Test Drivers' Cumulative Disengagement Rate

In Figure 5.2C, the Y-axes represent the cumulative disengagement rates. The data points are the cumulative disengagement rates at month-end for the month and year shown on the X-axis.

The right Y-axis and gray diamonds in Figure 5.2C represent the test drivers' cumulative disengagement rate. This rate (plotted at the end of each month) had a downward trend, as is self-evident by visual inspection but confirmed by a linear regression fit ($r = -0.70$, df = 12, $P = 0.005$).

5.3.5. The Effect of the Test Drivers' Cumulative Disengagement Rate on the AV Cumulative Crash Rate

The test drivers' cumulative disengagement rate was interpolated at each of the Google AV crash dates. Then the linear regression fit between the AV cumulative crash rate in Figure 5.2A and the interpolated test drivers' cumulative disengagement rate in Figure 5.2B had $r = -0.75$ (df = 4, $P = 0.09$). This result suggests that the upward trend in the Google AV cumulative crash rate for 2015–16 in Mountain View and vicinity (Figure 5.2A) is associated with a general downward trend in the test drivers' cumulative disengagement rate (Figure 5.2C, orange triangles). In other words, the upward trend in the AV cumulative crash rate may simply reflect a downward trend in the test drivers' cumulative disengagement rate. This result confirms and extends the 2015 result seen in Figure 2.2A and Figure 2.2C.

FIGURE 5.2 **Comparison of trends during the 14 month study period of Teoh and Kidd (2017) [5.1]. X-Axis**: Month and year from July 2015 to September 2016. **A. Y-Axis**: Google AV cumulative crash rate for "police-reportable" crashes in Mountain View, CA. **Symbols**: Six Google AV "police-reportable" crashes in Mountain View, CA at dates shown; see Figure 5.1 caption for symbol key. **B. Y-Axis**: Police-recorded CV cumulative crash rate for all police-recorded CV crashes in Mountain View, CA. **Symbols**: Y-Axis rate at quarter-end. **C. Left Y-Axis**: Google ADS cumulative disengagement rate. **Orange triangles**: Left Y-Axis rate at month-end. **Right Y-Axis**: Google AV test drivers' cumulative disengagement rate. **Gray diamonds:** Right Y-Axis rate at month-end. Dotted lines in the panels are linear regression fits.

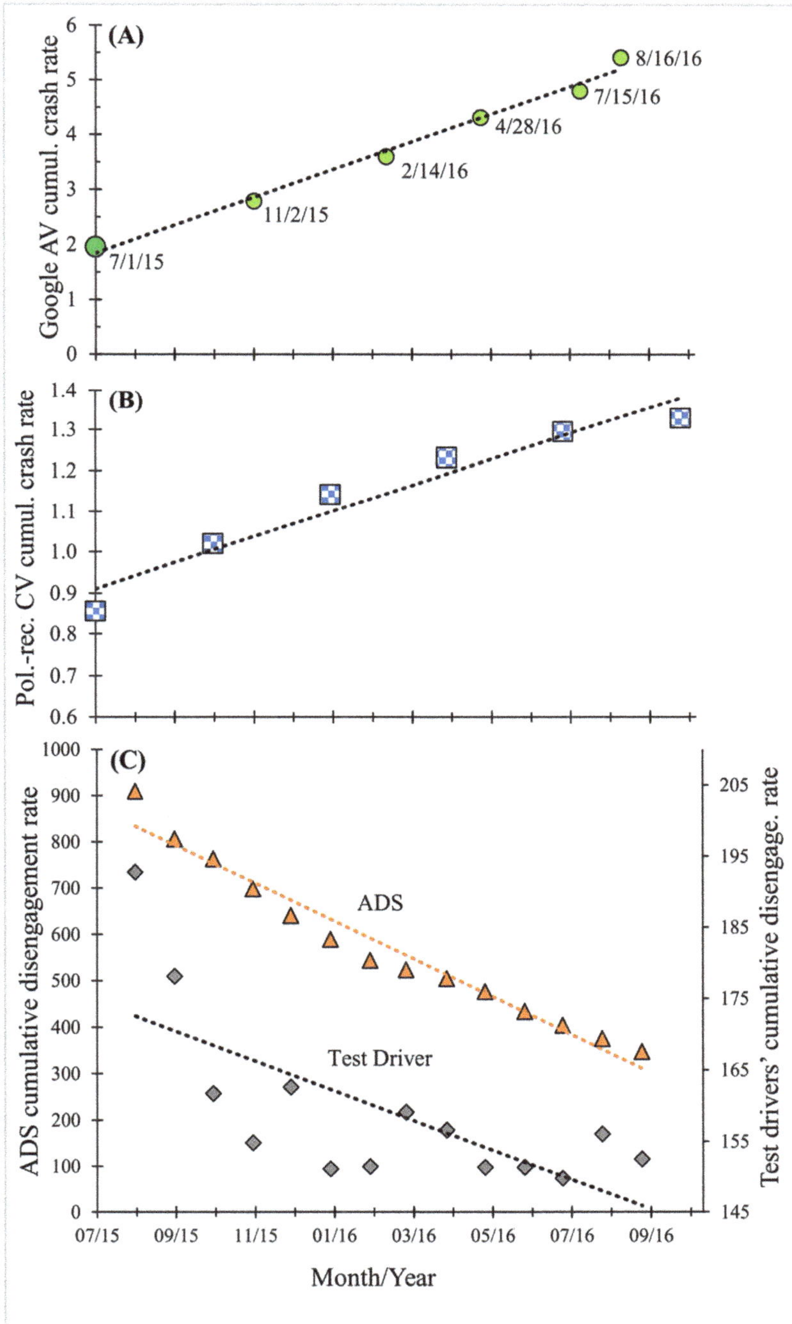

5.3.6. The Trend in the ADS Cumulative Disengagement Rate

In Figure 5.2C, the left Y-axis and orange triangles represent the ADS cumulative disengagement rate. Figure 5.2C extends Hypothesis 4 in Section 2.4.2.2.4 from test driver disengagements to ADS disengagements. The ADS cumulative disengagement rate also exhibits a downward trend as indicated by the linear regression fit ($r = -0.98$, df = 12, $P = 6.9E-12$).

5.3.7. The Effect of the ADS Cumulative Disengagement Rate on the AV Cumulative Crash Rate

The ADS cumulative disengagement rate was interpolated at the Google AV crash dates. The correlation coefficient for the regression fit between panel A (AV crash rate) and panel C (ADS Cumulative Disengagement Rate) was $r = -0.96$ (df = 4, $P = 0.002$). This result suggests that the upward trend in the Google AV cumulative crash rate during 2015 and 2016 (Figure 5.2A) may be explained by a general downward trend in the ADS cumulative disengagement rate (Figure 5.2C, orange triangles). In other words, decreasing test drivers' disengagement rates may increase AV crash rates.

5.4. Discussion of Reanalyzed Teoh and Kidd (2017) Data

Correction and adjustment of the original Teoh and Kidd (2017) [5.1] Phase 1 data and analysis gave rise to the partly-corrected data in Table 5.2. The revised Google AV crash IRR estimate of 2.11 (Table 5.2, note i) was now above one (causative) rather than below one (protective). This causative effect in the revised data is the opposite of the original Teoh and Kidd's (2017) [5.1] conclusion based on their original data in Phase 1 (Table 5.1). In other words, the revised Phase 1 data are inconsistent with the Phase 1 conclusion of Teoh and Kidd (2017) [5.1] that "Google self-driving cars…are safer than conventional human-driven passenger vehicles."

Furthermore, the continuing rise in the Google AV cumulative crash rate (Figure 5.2A) in 2016, as it did in 2015 (Chapter 2, Figure 2.2A), indicates that Google AVs became less safe during the Phase 2 study period of Teoh and Kidd (2017) [5.1]. This result is also inconsistent with Teoh and Kidd's (2017) [5.1] conclusion that "Google self-driving cars…are safer than conventional human-driven passenger vehicles."

Teoh and Kidd (2017, p. 60) [5.1], as noted in Section 5.2.1, conjectured that Google increased its AV usage on "higher-speed city streets" or had more "aggressive testing practices" in Phase 2 compared to Phase 1. No evidence to support those conjectures was presented by Teoh and Kidd (2017) [5.1]. (This hypothesis was labeled Hypothesis 1 in Section 2.4.2.2.1.)

Two alternative hypotheses were proposed and tested in Chapter 2 (see Hypotheses 3 and 4 in Sections 2.4.2.2.3 and 2.4.2.2.4) for the upward trend in the Google AV cumulative crash rate in 2015 (see Figure 2.2A). These hypotheses are tested further in this Chapter because Figure 5.2A indicates that the rise in the Google AV cumulative crash rate continued in 2016.

Hypothesis 3 states that the upward trend in the Google AV cumulative crash rate may be caused by the upward trend in the CV cumulative crash rate in Mountain View. (For the 2015 CV upward trend, see Figure 2.2B in Chapter 2; for the 2016 CV upward trend, see Figure 5.2B.) In other words, the upward trend in the cumulative crash rate of the Google AV fleet simply reflects a general upward trend in the probability of a crash (the CV cumulative crash rate) in Mountain View during 2015 and 2016.

Hypothesis 4 postulates a downward trend in the Google test drivers' cumulative disengagement rate (Figure 2.2C in Chapter 2 and Figure 5.2C, blue diamonds). This Chapter extends that hypothesis to include a downward trend in the Google ADS cumulative disengagement rate (Figure 5.2C, orange triangles). Hypothesis 4, therefore, states that a decreasing cumulative disengagement rate (either ADS-initiated or test-driver initiated) increased the Google AV cumulative crash rate in 2015 and 2016.

Hypotheses 3 and 4 are more parsimonious than Hypothesis 1 offered by Teoh and Kidd (2017) [5.1] (see Section 2.4.2.2.1), that Google changed its testing procedures. Hypotheses 3 and 4 are also more straightforward than Hypothesis 2 in Section 2.4.2.2.2 Hypothesis 2 postulated that the Google ADS neural networks (and the crash simulator methods for training it) exhibited negative learning (see Appendix M).

In short, the upward trend in the Google AV cumulative crash rate in 2015 and 2016 may have arisen from 1) an upward trend in the CV cumulative crash rate in Mountain View, 2) a decreasing trend in the ADS cumulative disengagement rate, 3) a decreasing trend in the test drivers' cumulative disengagement rate or any combination of those. Only additional AV and CV data can resolve which of these testable hypotheses, if any, are valid.

The hypotheses should be easy to falsify or validate with future data. For example, suppose future AV data show a downward trend in the AV cumulative crash rate and an upward trend in the CV cumulative crash rate in the same ODD and period. In that case, an AV developer can correctly claim that their AVs improve safety compared to CVs. (There seemed to be some evidence for that with Severe crash rates, as shown in Figure 3.3.) Hypothesis 3 will then be falsified.

Likewise, suppose a downward trend in the AV cumulative crash rate and an upward trend in the AV test drivers' cumulative disengagement rate in the same ODD and period (the opposite of Figure 5.2C). In that case, an AV developer company can also correctly claim that their AVs improve safety compared to CVs. Hypothesis 4 will then be falsified.

Furthermore, suppose a downward trend in the AV cumulative crash rate and an upward trend in the ADS disengagement rate in the same ODD and period. In that case, an AV developer can also correctly claim that their AVs are improving safety compared to CVs. The extension of Hypothesis 4 to ADS disengagements (Figure 5.2C, orange triangles) will then be falsified.

References and Endnotes

5.1. Teoh, E.R. and Kidd, D.G., "Rage Against the Machine? Google's Self-Driving Cars versus Human Drivers," *Journal of Safety Research* 63 (2017): 57–60, https://www.researchgate.net/profile/Eric_Teoh/publication/317265345, accessed May 19, 2021.

5.2. CA DMV, "Report of Traffic Collision Involving an Autonomous Vehicle (OL 316)," California Department of Motor Vehicles, 2014–2019, https://www.dmv.ca.gov/portal/uploads/2021/05/OL316.pdf, accessed August 14, 2021.

5.3. Google Reports, "Google Self-Driving Car Project Monthly Reports," Mountain View, CA, May 2015–November 2016. Note: Monthly reports are longer publicly available from Google. However, 19 monthly Google Reports from May 2015 to November 2016 are available at https://github.com/wgetsnaps/ca-dmv-autonomous-vehicle-accidents/tree/master/google-pdfs, accessed May 19, 2021.

5.4. HPMS, "California Department of Transportation Highway Performance Monitoring System Annual Reports 2009–2014," https://dot.ca.gov/programs/research-innovation-system-information/highway-performance-monitoring-system, accessed May 19, 2021.

5.5. SWITRS, "California Highway Patrol Statewide Integrated Traffic Records System," https://iswitrs.chp.ca.gov/Reports/jsp/index.jsp, accessed May 19, 2021.

5.6. StatsDirect, Statistical Software, Version 3.2.8, September 16, 2019, https://www.statsdirect.com/, accessed September 22, 2021.

5.7. CA DMV, "California Code of Regulations Title 13, Division 1, Chapter 1, Article 3.7—Testing of Autonomous Vehicles, §227.50," 2019, https://www.dmv.ca.gov/portal/uploads/2020/06/Adopted-Regulatory-Text-2019.pdf, accessed May 19, 2021.

5.8. Schoettle, B. and Sivak, M., "A Preliminary Analysis of Real-World Crashes Involving Self-Driving Vehicles," Technical Report No. UMTRI-2015-34, The University of Michigan Transportation Research Institute, Ann Arbor, MI, 2015, http://www.umich.edu/~umtriswt/PDF/UMTRI-2015-34.pdf, accessed May 19, 2021.

5.9. Hauer, E., "Engineering Judgment and Road Safety," *Accident Analysis & Prevention* 129 (2019): 180–189, doi:10.1016/j.aap.2019.04.022.

5.10. Blanco, M., Atwood, J., Russell, S., Trimble, T. et al., "Automated Vehicle Crash Rate Comparison Using Naturalistic Data," Virginia Tech Transportation Institute, January 2016, https://vtechworks.lib.vt.edu/handle/10919/64420, accessed May 19, 2021.

5.11. Dixit, V.V., Chand, S., and Nair, D.J., "Autonomous Vehicles: Disengagements, Accidents and Reaction Times," *PLOS ONE* 11, no. 12 (2016): e0168054, doi:10.1371/journal.pone.0168054, https://doi.org/10.1371/journal.pone.0168054, accessed May 19, 2021.

5.12. Google Disengagements, 2015, 2016, "Google Self-Driving Car Testing Report on Disengagements of Autonomous Mode," Google, Inc., Mountain View, CA, 2015, 2016, https://www.consumerwatchdog.org/resources/cadmvdisengagereport-dec.2015.pdf, accessed May 19, 2021. 2016 report available by request from: https://www.dmv.ca.gov/portal/vehicle-industry-services/autonomous-vehicles/disengagement-reports/.

5.13. Sivak, M. and Schoettle, B., "Road Safety with Self-Driving Vehicles: General Limitations and Road Sharing with Conventional Vehicles," Technical Report No. UMTRI-2015-02, The University of Michigan Transportation Research Institute, Ann Arbor, MI, 2015, http://deepblue.lib.umich.edu/handle/2027.42/111735, accessed May 19, 2021.

6

Heterogeneity and Crash Selection Bias: Favarò et al. (2017)

Contents

Note: Refer to Appendix B on page 149 for an explanation of the abbreviations and the definitions of unique terms and phrases.

The CA DMV regulations that required reporting AV crashes and AV mileage on public roads in CA went into effect on September 16, 2014. After that time, all companies wishing to test AVs on CA public roads had to file crash and mileage reports into a dedicated publicly accessible database at the CA DMV (separate from the police database for crashes). The Favarò et al. (2017) [6.1] study was the first to analyze only the required OL-316 AV crash reports (CA DMV, 2014–2019) [6.2] and the required disengagement reports which contain the AV mileage (CA DMV, 2015–2018) [6.3]. The four prior studies, reviewed in Chapters 2, 3, 4, and 5 of this book, used only the Google Reports [6.4], which contained Google AV crash and mileage data from 2009–2015.

6.1. Summary of Favarò et al. (2017) Original Data

6.1.1. Original Data in Tabular Format

Favarò et al. (2017, Table 3) [6.1] analyzed 26 OL-316 crash reports [6.2] from September 1, 2014, to March 30, 2017. Table 6.1 shows that two of the crashes were from Nissan vehicles (one filed by Nissan and the second filed on January 8, 2016, by Cruise Automation using a Nissan Leaf), one from a Delphi/Audi vehicle, one from a GM Cruise Chevrolet, ten from a Google prototype AV with a maximum speed of 25 mph, and 12 from a Lexus vehicle retrofitted with a Google ADS.

Favarò et al. (2017) [6.1] combined across the AV crash severities within each of the five vehicle types to obtain the AV crash counts shown in Table 6.1. The AV MVMT is the total AV mileage for the entire fleet of each vehicle type for the study period.

Favarò et al. (2017) [6.1] then combined the AV crash and AV MVMT data across the five vehicle types to obtain totals of 26 AV crashes and 1.088453 MVMT (Table 6.1, notes b and c). Thus, the total AV crash rate estimate was 23.9 AV crashes per MVMT (Table 6.1, note d). This estimate is not made **bold** in Table 6.1 because it has some biases and best practice violations, as described in the later sections of this chapter.

Favarò et al. (2017) [6.1] next compared their AV total crash rate estimate of 23.9 (Table 6.1, note d) to the total 2015 U.S. CV total crash rate estimate of 2.00 (Table 6.1, note h).

TABLE 6.1 Favarò et al. (2017): Original data.[a]

Vehicle type	AV crashes	AV MVMT	AV crash rate[a]
Nissan (Nissan & GM Cruise)	**2**	**0.007152**	**279.6**
Delphi/Audi	**1**	**0.019787**	**50.5**
Chevrolet (GM Cruise)	**1**	**0.008447**	**118.4**
Google prototype	**10**	**0.403226**	**24.8**
Google retrofitted Lexus	**12**	**0.649841**	**18.5**
Total AV	**26**[b]	**1.088453**[c]	23.9[d]
Total CV[e]	**6,296,000**[f]	3,148,000[g]	2.00[h]

Favarò et al.'s (2017) [6.1] data for five vehicle types in CA AV tests from September 1, 2014, to March 30, 2017. Data from Favarò et al. (2017, Tables 3, 4) [6.1].

A **bold** number in the Table above indicates no identified error or bias.

Underlined footnote words below are revised in later tables.

Notes:

[a] Favarò et al.'s (2017) [6.1] original variable names are translated to be consistent with the terminology in this book. For example, "crash frequency" in Favarò et al. (2017, Table 3) is here termed "crash rate." The crash rate is estimated as crashes ÷ MVMT.

[b] Total AV crashes combined across vehicle types.

[c] Total AV MVMT combined across vehicle types.

[d] Crude total AV crash rate.

[e] Preliminary 2015 U.S. national CV data estimates.

[f] Preliminary total CV crashes.

[g] Preliminary total CV MVMT.

[h] Preliminary CV total crash rate estimate.

Favarò et al. (2017) [6.1, p. 17] concluded that there was "…one order of magnitude difference between AVs and conventional vehicles for…accident frequency." In other words, Favarò et al.'s (2017) [6.1] conclusion referred to their result that their total AV total crash rate estimate was more than ten times larger than their CV total crash rate estimate in their original data, as reproduced in Table 6.1.

6.1.2. Original Total AV and CV Data in Epidemiologic Format

Table 6.2 puts Favarò et al.'s (2017, Table 4) [6.1] Total AV and Total CV data in Table 6.1 into the epidemiologic format used for comparing two rates (see Chapter 1, Section 1.1, and Table 1.1). The epidemiologic design in Table 6.2 makes it easier to notice biases and errors. It also puts the data into a standard format for easier comparison across studies.

Table 6.2 compares the AV total crash rate of 23.9 (Table 6.2, note d) to the final 2015 U.S. CV total crash rate of 2.03 (Table 6,2, note e). The resulting AV crash IRR estimate of 11.9 (Table 6.2, note i) indicates that the AV total crash rate estimate is an order of magnitude increase compared to the 2015 U.S. CV total crash rate. This epidemiologic result based on the IRR metric is consistent with Favarò et al.'s (2017) [6.1, p. 17] conclusion based on crash frequencies that there was "one order of magnitude difference between AVs and conventional vehicles." However, as shown in the next section, this IRR estimate of 11.9 contains two best practice violations and a substantial unrecorded crash selection bias.

TABLE 6.2 Epidemiologic analysis of total data of Favarò et al. (2017, Table 4).

	Total AV[a]	2015 U.S. CV[e]
Total crashes	**26**[b]	**6,296,000**[f]
Total MVMT	**1.088453**[c]	**3,095,000**[g]
Crash rate	23.9[d]	**2.03**[h]
IRR (95% CI)[i]	11.9 (7.8–17.0), *P* = 3.0E–19	

Favarò et al.'s (2017) [6.1] AV total crash rate across all vehicle types from Table 6.1 is compared to the final 2015 U.S. police-recorded CV total crash rate. The AV crash IRR and its 95% confidence limits are estimated using the methods in Chapter 1.

A **bold** number in the 2×2 data matrix above indicates no identified error or bias.

Underlined footnote words below are not final.

Notes:

[a] Totals for five AV types in CA AV testing from September 1, 2014, to March 30, 2017.

[b] Total AV crashes. Combined AV crashes across vehicle types (Table 6.1, note b).

[c] Total AV mileage across vehicle types (Table 6.1, note c).

[d] AV total crash rate estimate. Not final for reasons to be discussed in Section 6.2.

[e] Final 2015 U.S. national CV data (National Center for Statistics and Analysis, 2020) [6.5].

[f] Final 2015 police-recorded CV total crash estimate [6.5].

[g] Final 2015 CV Total MVMT estimate on all city streets, roads, and highways in the U. S. [6.5].

[h] Final CV crash rate estimate.

[i] The exact mid-P confidence levels and two-tailed P of the IRR were estimated with the StatsDirect rate ratio program [6.6].

6.2. Critical Analysis of Favarò et al. (2017)

6.2.1. Best Practice Violation 1: Heterogeneous Vehicle Types (Table 6.2, Note h)

The AV crash rate estimates for the five vehicle types in Table 6.1 were checked for heterogeneity. The lower and upper confidence limits (not shown) for each of the five AV crash rates in Table 6.1 were calculated using the epiTab program in Episheet (Rothman, 2015) [6.7]. The five crash rates with their confidence limits were then tested for heterogeneity with the meta-analysis program in Episheet (Rothman, 2015) [6.7]. The five AV crash rates in Table 6.1 are heterogeneous ($P = 0.0004$). The heterogeneity is not surprising, given the wide range of AV crash rates in Table 6.1, from 18.5 to 279.6. Hence, it was a heterogeneity best practice violation (see Chapter 1, Section 1.3) for Favarò et al. (2017) [6.1] to estimate a total AV total crash rate by combining the five vehicle types in Table 6.1. The crash rates of the vehicle types are heterogeneous, and therefore it is a best practice violation to combine them into an overall crash rate. The best practice (see Chapter 1, Section 1.3) stratifies the data by the five vehicle types and analyzes each vehicle type separately.

6.2.2. Best Practice Violation 2: Heterogeneous CV Sites (Table 6.2, Notes b and f)

Favarò et al.'s (2017) [6.1] analysis method contained a second heterogeneity best practice violation because the AV types they analyzed had different ODDs with differing CV crash rates. For example, GM Cruise AVs were tested on San Francisco city streets, with higher CV crash rates than the national average (Appendix K, Table K.3). Delphi/Audi AV types were tested on highways across the United States, with lower CV non-fatal crash rates than urban driving (see Chapter 2, Section 2.1.3). Google AV types were tested on city streets in Mountain View and vicinity with lower CV crash rates than the national average (Appendix K, Table K.1).

Because of heterogeneous CV sites, none of the sites analyzed by Favarò et al. (2017) [6.1] in Table 6.1, whether singly or summed, likely had a CV crash rate representative of the national average CV crash rate. The national CV crash rate is not just on city streets but includes heavy and light vehicles, different weather conditions, etc. Hence, any substantial differences between an AV crash rate and the national CV crash rate in Table 6.1 may simply reflect differences in the CV crash rates at each AV test site compared to the national CV site. It is plausible that sites with higher CV crash rates would have higher AV crash rates than sites with lower CV crash rates, simply because of the many site factors that can affect the probability of a crash, whether an AV or CV crash (see Chapter 3, Section 3.3). The result is that any AV tested at a given site would have its AV crash data biased by the differences between the CV crash rate at that site and the national average. An IRR estimate based on the total AV data of Favarò et al. (2017) [6.1] in Table 6.1 at many sites, such as Table 6.2, would also confound any AV versus CV IRR effect size with site selection bias.

The best practice is to stratify each data by each site (see Chapter 1, Section 1.3) and not attempt an overall analysis as in Table 6.2. Also, each site should have its AV crash data standardized to control for the crash rate difference between CVs at the AV test site and

CVs at the comparison site. Chapter 3 (Section 3.4) and Appendix H illustrate the site standardization method that compares the CV crash rates at different sites to control for modification of the effect size of the AV crash IRR estimate by site selection bias.

6.2.3. Unrecorded Crash Selection Bias (Table 6.2, Note f)

The analysis of the original data presented in the epidemiologic format of Table 6.2 found that Favarò et al.'s (2017) [6.1] methods contained an unrecorded crash selection bias in addition to its best practice violations. The epidemiologic method compared the AV estimate of 23.9 (Table 6.2, note d) to the 2015 U.S. national CV total crash rate estimate of 2.03 (Table 6.2, note f). However, every AV crash rate estimate in Table 6.1 included all AV crashes, both police-recorded and police-unrecorded. However, the 2015 U.S. average CV rate of 2.03 crashes per MVMT (Table 6.2, note h) includes police-recorded crashes only. Therefore, the AV IRR estimate of 11.9 (Table 6.2, note i) is biased substantially upward. The main reason is unrecorded crash selection bias arising from Favarò et al.'s (2017) [6.1] differential selection criterion for AV crashes (police-recorded and police-unrecorded crashes) versus CV crashes (police-recorded crashes only).

This unrecorded crash selection bias, if not adjusted for, causes a substantial overestimate of the AV crash IRR. For example, Appendix E (Table E.1, bottom row) shows that the VTTI SHRP2 NDS had 95% CV crashes without a Police Accident Report (PAR). The investigators were able to count every CV crash that occurred because all the vehicles were heavily instrumented. (AVs are also heavily instrumented, so every AV crash is also counted.) The VTTI investigators thoroughly investigated whether every crash in the SHRP2 database had a PAR or not. They found that 95% of the known SHRP2 CV crashes had no PAR (see Chapter 3). Some SHRP2 crashes without a PAR were even Severe crashes with injuries and towaways (see Appendix E). So there would be an automatic but false IRR estimate of 20 for comparing AV crashes to CV crashes with a PAR. But not all crashes with a PAR have a record in the state police databases. (For example, Minor PDO crashes may have a PAR, but would not be entered into state police databases, which generally have only more serious crashes in them.) If only 1 CV crash out of 100 is finally entered into a state police database to become a police-recorded crash, the AV crash IRR would be falsely estimated as 100.

6.3. Discussion of Favarò et al. (2017)

6.3.1. Corrections Not Made in Favarò et al.'s (2017) Data

As with the Dixit et al. (2016) [6.8] study (Chapter 4, Section 4.2), no corrections could be made for the Favarò et al.'s (2017) [6.1] data in Table 6.2. The reason is that Favarò et al. (2017) [6.1] did not give sufficient information to enable correcting their best practice violations of combining heterogeneous companies (Section 6.2.1) and heterogeneous CV sites (Section 6.2.2). In addition, there was no easy way to adjust for Favarò et al.'s (2017) [6.1] unrecorded crash selection bias (Section 6.2.3).

6.3.2. Favarò et al. (2018) Study

Favarò et al. (2018) [6.9] published a subsequent study of AV disengagements from 2014 to 2017 using methods similar to those in Favarò et al. (2017) [6.1]. Favarò et al. (2018) [6.9] again had a small section on crashes, using a similar analysis method to Favarò et al. (2017) [6.1]. Favarò et al. (2018, Figure 10) [6.9] continued their best practice violation by summing heterogeneous companies (Section 6.2.1) in their crash data analyses. In addition, they again had unrecorded crash selection bias (Section 6.2.3) when comparing AV crash frequencies to the 2015 U.S. national CV crash frequency. Unfortunately, no crash counts or crash mileages were published in their article, so their best practice violations and biases could not be corrected using epidemiologic methods, as was done in Chapters 2, 3, 4, and 5. Therefore, further analysis was discontinued of the data in Favarò et al. (2018) [6.9] as well as Favarò et al. (2017) [6.1]. Because of these uncorrected issues, the estimated AV crash IRR of 11.9 based on Favarò et al.'s (2017) [6.1] data in Table 6.2 should not be considered valid.

References and Endnotes

6.1. Favarò, F.M., Nader, N., Eurich, S.O., Tripp, M. et al., "Examining Accident Reports Involving Autonomous Vehicles in California," *PLOS ONE* 12, no. 9 (2017): e0184952, doi:10.1371/journal.pone.0184952, http://journals.plos.org/plosone/article?id=10.1371/journal.pone.0184952, accessed May 19, 2021.

6.2. CA DMV, "Report of Traffic Collision Involving an Autonomous Vehicle (OL 316)," California Department of Motor Vehicles, 2014–2019, https://www.dmv.ca.gov/portal/uploads/2021/05/OL316.pdf, accessed August 14, 2021.

6.3. CA DMV, "Disengagement Reports," 2015–2018, https://www.dmv.ca.gov/portal/vehicle-industry-services/autonomous-vehicles/disengagement-reports/. Note: Available upon request. Email to AVarchive@dmv.ca.gov to request a digital copy of an archived report. Requests must include the year and manufacturer.

6.4. Google Reports, "Google Self-Driving Car Project Monthly Reports," Mountain View, CA, May 2015–November 2016. Note: Monthly reports are longer publicly available from Google. However, 19 monthly Google Reports from May 2015 to November 2016 are available at https://github.com/wgetsnaps/ca-dmv-autonomous-vehicle-accidents/tree/master/google-pdfs, accessed May 19, 2021.

6.5. National Center for Statistics and Analysis, "Summary of Motor Vehicle Crashes (Final Edition): 2015 Data," Traffic Safety Facts, Report No. DOT HS 812 376, National Highway Traffic Safety Administration, Washington, DC, February 2020, https://crashstats.nhtsa.dot.gov/Api/Public/ViewPublication/812376, accessed June 2, 2021.

6.6. StatsDirect, Statistical Software, Version 3.2.8, September 16, 2019, https://www.statsdirect.com/, accessed September 22, 2021.

6.7. Rothman, K.J., "Episheet: Spreadsheets for the Analysis of Epidemiologic Data," 2015, https://www.colleaga.org/tools/episheet, accessed May 19, 2021.

6.8. Dixit, V.V., Chand, S., and Nair, D.J., "Autonomous Vehicles: Disengagements, Accidents and Reaction Times," *PLOS ONE* 11, no. 12 (2016): e0168054, doi:10.1371/journal.pone.0168054, https://doi.org/10.1371/journal.pone.0168054, accessed May 19, 2021.

6.9. Favarò, F., Eurich, S., and Nader, N., "Autonomous Vehicles' Disengagements: Trends, Triggers, and Regulatory Limitations," *Accident Analysis & Prevention* 110 (2018): 136–148.

7

Crash Selection Bias and Site Selection Bias: Banerjee et al. (2018)

Contents

Note: Refer to Appendix B on page 149 for an explanation of the abbreviations and the definitions of unique terms and phrases.

Banerjee et al. (2018) [7.1], computer scientists and engineers at the University of Illinois at Urbana-Champaign, used "failure analysis" methods to analyze AV data. They analyzed CA DMV AV disengagement reports [7.2] and CA DMV AV OL-316 crash reports [7.3] for 26 months from September 2014 to November 2016. Banerjee et al.'s (2018) [7.1] study was the second one to analyze the CA DMV crash and disengagement reports solely, after the Favarò et al. (2017) [7.4] study that was reviewed in Chapter 6. Earlier studies (Chapters 2, 3, 4 and 5) examined the early disengagement and crash reports published directly by Google.

7.1. Summary of Banerjee et al. (2018)

The main focus of Banerjee et al. (2018) [7.1] was disengagements, like the Dixit et al. (2016) [7.5] study reviewed in Chapter 4. Disengagements are outside the main scope of this book. This book focuses on AV crash rates and their comparison to CV crash rates. However, this book does a limited analysis of disengagements regarding their possible effects on AV crash rates (see Figures 2.2C and 5.2C). Nonetheless, Banerjee et al. (2018) [7.1] devoted a small portion of their article to crashes. See their Section V.B in "Analysis of AV Accident Reports" (pp. 594–595). They examined the CA DMV AV crash reports [7.3] from September 1, 2014, to November 30, 2016, to tabulate AV crashes ([7.1], Table VI). They also tabulated the AV monthly mileages over the same period using the CA DMV disengagement reports [7.2].

Banerjee et al. (2018) [7.1] then developed a unique new metric for estimating AV crash rates for AV manufacturers from these crash and mileage data, which they called Median APM, or "median accidents per mile per vehicle." (Their method for estimating this metric is described in more detail in the following section.) Four of the 12 AV manufacturers had more than zero AV test miles (Waymo/Google, Delphi, Nissan, and GM Cruise). Hence, these investigators attempted to estimate a Median APM for these four AV manufacturers ([7.1], Table VII). They then compared each Median APM to the preliminary 2015 U.S. national CV crash rate ([7.1], Table VII), which they called HAPM or Human Accidents per Mile.

Table 7.1 shows the data (columns 1 and 2) and metrics for these four companies. The numbers directly from Banerjee et al.'s (2018) [7.1] article are in **bold**. This review added the metrics in non-bold.

Banerjee et al. (2018) [7.1] compared their "3. Median APM" estimate for each AV manufacturer to the CV crash rate estimate labeled "5. HAPM" in Table 7.1. The investigators used the preliminary 2015 estimate from NHTSA (2016) [7.6], or 2.00 crashes per MVMT, for their human CV crash rate estimate. NHTSA (2016) [7.6] estimated this U.S. national CV crash rate from state summaries of mileage and police-reported CV crashes. This estimate was an early release before the final 2015 U.S. CV estimate as used in Chapter 6 in the reanalysis (Table 6.2) of the Favarò et al. (2017) [7.4] data.

Banerjee et al. (2018) [7.1] then divided their "3. Median APM" AV crash rate estimates by the "5.0 HAPM," CV crash rate estimate to calculate their metric in the "6. Rel. to HAPM" column. Banerjee et al. (2018) [7.1] state that their metric for the relative rates of AVs to CVs ranged from a lowest (best) with Nissan (a Median APM rate that was 15.285 times worse than the national CV average in their original data). Waymo was next worst at 20.7 times and then Delphi at a 22.99 Median APM ratio compared to CVs. The worst was GM Cruise, with a relative Median APM rate of 4421.5 times the national CV average. Banerjee et al. (2018, Abstract) [7.1] conclude, "...autonomous vehicles are 15 – 4000× worse than human drivers for accidents per cumulative mile driven."

After correcting the typo for the Nissan "6. Rel. to HAPM" metric, their conclusion would read "...autonomous vehicles are 21 – 4000× worse than human drivers for accidents per cumulative mile driven."

Banerjee et al.'s (2018) Method to Estimate Median APM

Banerjee et al. (2018) [7.1] attempted to applied engineering reliability metrics and methods to analyze AV failure rates. On the face of it, this method could have some advantages over

TABLE 7.1 Banerjee et al.'s (2018) data are summarized in **bold** and the metric is in *italics* in columns 1–3 and 5–6. The non-bold and non-italic data in columns 4 and 7–9 were estimated in this review for comparison purposes.

Manufacturer[a]	1. *Accidents*[b]	2. *Miles*[c]	3. *Median APM*[d,e]	4. *Median APM*[d] × 10^6	5. *HAPM*[f]	6. *Rel. to HAPM*[g]	7. AV crash rate[h]	8. AV crash IRR[i]	9. % Diff. Vars. 6 and 8[j]
Waymo	**25**	**1,060,200**	$\mathbf{4.140 \times 10^{-5}}$	41.4	**2.0**	**20.7**	23.6	11.8	76%
Delphi	**1**	**19,751**	$\mathbf{4.599 \times 10^{-5}}$	46.0	**2.0**	**22.99**	50.6	25.3	−9%
Nissan	**1**	**5,584.4**	$\mathbf{3.057 \times 10^{-4}}$	305.7	**2.0**	152.85[k]	179.1	89.54	71%[l]
GM Cruise	**14**	**10,015.2**	$\mathbf{8.843 \times 10^{-3}}$	8843.0	**2.0**	**4421.5**	1397.9	689.9	533%

Summary of Banerjee et al.'s (2018) [7.1] original data in their Tables I and VI–VIII for four AV manufacturers in CA prototype AV tests from 2016 to 2017.

Bold entries in the table above are directly quoted from Banerjee et al. (2018) [7.1].

Notes:

[a] AV manufacturer.

[b] *Accidents* = AV crashes. From [7.1, Tables I or VI].

[c] *Miles* = AV miles. From [7.1, Table I].

[d] *Median APM*. From [7.1, Table VII].

[e] *APM* = Accidents per mile (same as crash rate).

[f] *HAPM* = *Human APM* = 2015 preliminary national CV crash rate from NHTSA (2016) [7.6]. [7.1, Table VII, footnote].

[g] *Rel. to HAPM* = (4. *Median APM*[d] ÷ 5. *HAPM*[e]). [7.1, Table VII, last column].

[h] AV crash rate = 10^6 × (1. *Accidents*[b] ÷ 2. *Miles*[c]).

[i] AV crash IRR = (7. AV crash rate[h] ÷ 5. *HAPM*[f]).

[j] % Difference in Variables 6 and 8 = 100 × (6. *Rel. to HAPM*[g] - 8. AV crash IRR[i]) ÷ (8. AV crash IRR[i]).

[k] Corrected typo in Banerjee et al. (2018, Table VII) [7.1] for Nissan for metric "6. *Rel. to HAPM*." The original estimate was 15.285. It was corrected to 152.58, shown in non-bold.

[l] After correcting the Nissan data typo described in note k, the "9. % Difference" estimate for Nissan changed from 83% to 71%.

standard metrics because using the total crashes and total miles for a manufacturer to estimate crash rates might overlook essential properties of the statistical distributions, which are well known to be non-normal. (Epidemiological metrics also recognize this non-normality for rare events, so they use Poisson rather than normal distributions.)

The direct method to estimate a failure rate (i.e., crash rate) for an AV manufacturer is to evaluate the failure rate for each vehicle in the manufacturer's AV fleet (i.e., estimate the APM per vehicle). Then one could attempt to estimate a median APM (or other reliability metrics) for a manufacturer's entire AV fleet from the statistical distribution of the individual vehicle APMs.

However, Banerjee et al. (2018, p. 594) [7.1] correctly noted, "…we cannot compute the APM per vehicle directly." The reason is that the CA DMV crash reports [7.3] give the crash date, but they do not provide a specific ID (name, VIN, etc.) of each vehicle used by the manufacturer. Unfortunately, the ID of a vehicle in a crash report [7.3] is required to match that vehicle ID for the crash to the vehicle ID in the monthly disengagement reports [7.2] to obtain the mileage for that vehicle. In addition, the CA DMV AV crash reports [7.3] do not give vehicle IDs for the AVs for unknown reasons. All the AVs are manufacturer-owned, so there is no personal identifying information that could be compromised. In addition, not every AV manufacturer supplies vehicle IDs in its disengagement reports [7.2]. Hence, a vehicle's mileage and disengagement data cannot be matched at a vehicle level to its crash data using any data in the DMV AV databases [7.2, 7.3]. Without matched vehicle IDs between the crashes and the mileage data, it is impossible to determine the APM of an individual vehicle in a given fleet.

Banerjee et al. (2018, bottom-left column of p. 594) [7.1] correctly recognized this problem and stated, "We instead compute accidents per mile using the equation APM = DPM/DPA." Equation (7.1) shows the elementary algebra.

$$APM = \frac{DPM}{DPA} = \frac{\frac{\text{Disengagements}}{\text{Mile}}}{\frac{\text{Disengagements}}{\text{Accident}}} = \frac{\text{Accidents}}{\text{Mile}} \tag{7.1}$$

7.2. Critical Analysis of Banerjee et al. (2018)

First, Banerjee et al. (2018) [7.1] had a minor error in the estimate of the Nissan relative crash rate of 15.285 ("6. Rel. to HAPM," data row 3). Second, there were anomalies between their data and figures concerning their metrics of disengagements per mile (DPM) and APM that could not be replicated. Third, there were substantial upward biases in the APM and Rel. to HAPM metrics as contained in the analysis methods used by Banerjee et al. (2018) [7.1].

7.2.1. Minor Error in Nissan Median APM

There was a typo in Banerjee et al. (2018, Table VII) [7.1] for the Nissan estimate in the "6. Rel. to HAPM" variable. That ratio was reestimated as ("4. Median APM × 10^6" ÷ "5. HAPM"), or (305.7 ÷ 2). The ratio equals 152.85, not the original 15.258 estimate of Banerjee et al. (2018, Table VII) [7.1]. See Table 7.1 (note j) above. After correction of the Nissan "6. Rel. to HAPM" typo, the % Difference between Banerjee et al.'s (2018, Table VII) [7.1] "6. Rel. to HAPM" estimate and the "8. AV crash IRR" estimate changed from −83% to 71%, as shown in Table 7.1.

7.2.2. Data Anomalies

One substantial data anomaly is that the estimates of the Median DPMs ("Median Disengagements per Mile") in Table VII in Banerjee et al. (2018) [7.1] do not always match the Median DPMs illustrated in their Figure 4 in the same article. For example, in Table VII, their Median DPM estimates show that GM Cruise has a Median DPM of 0.177 and Delphi of 0.0263, a sixfold higher Median DPM for GM Cruise than Delphi. However, Delphi and GM Cruise have almost identical Median DPMs in Figure 4 in Banerjee et al. (2018) [7.1]. (The Median DPM in Figure 4 for each manufacturer is the horizontal bar in each box.) The Delphi Median DPM in their Figure 4 *y*-axis is near the GM Cruise estimate (0.177) when it should be much lower (0.0263), according to the data in Table VII in Banerjee et al. (2018) [7.1].

7.2.3. Failure to Replicate Median DPM

To estimate a Median APM using Equation (7.1) requires estimating a Median DPM in the numerator of Equation (7.1). However, estimating the median of a vehicle DPM distribution first requires estimating the DPM for each vehicle in a given AV fleet. In turn, estimating the DPM at this vehicle level requires that AV disengagements and AV miles be available at the vehicle level. In other words, two metrics ("disengagements per vehicle" and "miles per vehicle)" are needed to estimate the DPM for an individual vehicle.

The symbols in Figure 4 in Banerjee et al. (2018) [7.1] indicate the study's estimates of such a DPM per vehicle per each AV manufacturer. The box-and-whisker chart in their study's Figure 4 shows the Median DPM per manufacturer as the horizontal line in each box, estimated from the distribution of the DPM symbols for all the vehicles for each manufacturer's AV fleet. Table VII in Banerjee et al. (2018) [7.1] gives the Median DPM estimates for eight AV manufacturers. Banerjee et al. (2018) [7.1] say they estimated the Median APMs for each manufacturer's AV fleet from these and the DPA estimates for each manufacturer. Thus, the Median APM estimates they made in their Table VII are reproduced here in Table 7.1 as "3. Median APM" for four of the AV manufacturers.

That Median DPM is a critical metric for Banerjee et al. (2018) [7.1] because it is what they say these used as per Equation (7.1) to estimate their final crash metric of Median APM. Therefore, to resolve the data anomaly in Section 7.2.2, an attempt was made to determine whether Banerjee et al.'s Figure 4 or Table VII represents the correct value for the Median DPM metric for each manufacturer (reproduced here in Table 7.1, column 3). Specifically, a reanalysis attempted to replicate the critical Median DPM metric using the CA DMV data [7.2, 7.3] cited by Banerjee et al. (2018) [7.1].

The critical variable Median DPM could not be replicated from the methods described by Banerjee et al. (2018) [7.1] from the data submitted by the AV manufacturers and published by the CA DMV [7.2, 7.3] for 2015–2016. Table 7.2 illustrates a possible reason for this failure to replicate.

Table 7.2A summarizes whether the required metrics of "Miles per Vehicle" and "Disengagements per Vehicle" were available in the CA DMV disengagement reports [7.2]. The 2016 disengagement reports [7.2] included data from 2014 to November 30, 2015. The 2017 disengagement reports [7.2] included data from December 1, 2015, to November 30, 2016, for the AV manufacturers shown. As noted, both metrics are required to estimate the Median DPM values per vehicle plotted as data points in Figure 4 in Banerjee et al. (2018) [7.1].

Table 7.2B summarizes what is known from Table 7.2A. Only Delphi provided data for both "Miles per Vehicle" and "Disengagements per Vehicle" for the period in the scope of Banerjee et al.'s (2018) [7.1] analysis. Therefore, Delphi is the only manufacturer for which the two datasets are available that are necessary to estimate DPM at the vehicle level, required to form a Median DPM in the numerator of Equation (7.1) used by Banerjee et al. (2018) [7.1].

Table 7.2A further shows that Nissan did not provide "Miles per Vehicle" in their CA DMV disengagement reports [7.2] for the 2014–2016 study period of Banerjee et al. (2018) [7.1]. In addition, Table 7.2A shows that Waymo and GM Cruise did not provide "Disengagements per Vehicle" in their CA DMV disengagement reports [7.2] for the 2014–2016 study period. Therefore, Table 7.2B shows that it was impossible to replicate the "DPM per vehicle" for Waymo, Nissan, and GM Cruise from the data those manufacturers submitted to the CA DMV in their disengagement reports [7.2]. Without vehicle-level DPM data, it was impossible to estimate a Median DPM for a manufacturer's AV fleet. Therefore, it was unclear how to replicate Banerjee et al.'s (2018) [7.1] vehicle-level Median DPM metric for any manufacturer except Delphi. In other words, without the metrics of "Miles per Vehicle" and "Disengagements per Vehicle," the "DPM per Vehicle" metric cannot be directly estimated for Waymo, Nissan, or GM Cruise. Therefore the Median DPM metric could not be replicated for those manufacturers.

TABLE 7.2 Vehicle-level data available in CA DMV disengagement reports [7.2] for mileage and disengagements from 2014 to November 30, 2016.

	AV Manufacturer			
	Waymo	Delphi	Nissan	GM Cruise
A. Are there vehicle-level data in the CA DMV disengagement reports [7.2] for the study period?				
Miles per Vehicle	Yes[a,b]	Yes[c,d]	No	Yes[g]
Disengagements per Vehicle	No	Yes[c,d]	Yes[e,f]	No
B. Is it possible to estimate DPA (Disengagements per Accident per vehicle) from the disengagement reports [7.2] for the study period?	No	Yes	No	No

Notes

[a] Google, "Google Self-Driving Car Testing Report on Disengagements of Autonomous Mode, Appendix B. Autonomous miles on public roads in California for each car and month," December 2015 [7.2]. Data from September 2014 to November 2015.

[b] Waymo, "Report on Autonomous Mode Disengagements for Waymo Self-Driving Vehicles in California, Appendix B. Autonomous miles on public roads in California for each car and month," January 5, 2017 (revised) [7.2]. Data from December 2015 to November 2016.

[c] Delphi Corporation, "Summary of Autonomous Vehicle Disengagements," January 1, 2016 [7.2]. Data from October 2014 to November 2015.

[d] Delphi Corporation, "Summary of Autonomous Vehicle Disengagements," January 1, 2017 [7.2]. Data from December 2015 to November 2016.

[e] Nissan, "Nissan Autonomous Vehicle Disengagement Report 2015, Results of Planned Tests," December 23, 2015 [7.2].

[f] Nissan, "Nissan Autonomous Vehicle Disengagement Report 2016, Results of Planned Tests," December 13, 2016 [7.2].

[g] GM Cruise, "Autonomous Vehicle Tester Program Annual Disengagement Report." Table: Autonomous miles per month per vehicle. December 22, 2016 [7.2]. Monthly autonomous miles per month per vehicle from June 2015 to November 2016.

In sum, Table 7.2 indicates that there appeared to be insufficient information available from three AV manufacturers in their CA DMV disengagement reports [7.2] for this review to replicate Banerjee et al.'s (2018) [7.1] vehicle-level DPM. Therefore, it was unclear how to replicate Banerjee et al.'s (2018) [7.1, Table VII] Median DPM estimates. Consequently, it is unclear how Banerjee et al. (2018) [7.1] could plot in that study's Figure 4 the DPM for individual vehicles for the Waymo, Nissan, and GM Cruise fleets, which are required to estimate a Median DPM for those AV manufacturers. In turn, the Median DPM was needed in Equation (7.1) to estimate the Median APM for each manufacturer.

7.2.4. Terminology Definition Questions

In response to an email question of how to reproduce the DPM for individual vehicle data in Figure 4 in their study [7.1], Banerjee [7.7] kindly sent a reply that noted the following. The computer code to produce all the graphs and tables used in Banerjee et al.'s (2018) [7.1] analysis is open source [7.8]. In addition, the computer code to calculate the metrics [7.9] and the code to plot the figures/visualizations is also publicly available [7.10]. In particular, Banerjee [7.7] kindly noted that the code for their Figure 4 plot is in Box "14" in [7.10]. The data frame "fpm_data" created in Box "10" in [7.10] corresponds to the data presented in their tables. However, after examining their source code, it was still unclear how to replicate the data plotted in Figure 4 of Banerjee et al. (2018) [7.1]. (To be fair, the author is not that familiar with the computer language they used in their programming, nor how to activate their source code and run their programs.)

Banerjee [7.7] further noted that when the metric of Disengagements per Vehicle was missing for a particular manufacturer (e.g., as per Table 7.2A above), the investigators substituted other metrics. For example, their study used an "Accident per cumulative mile metric," which aggregates miles across all vehicles driven by a manufacturer. However, it was still unclear which metrics were used for each manufacturer for their estimates of "3. Median APM," reproduced in Table 7.1 above. Thus, there are multiple possibilities in applying Equation (7.1) to estimate APM. Banerjee et al. (2018) [7.1] perhaps defined the terms in Equation (7.1) in different ways for different manufacturers, depending on the particular data available for each manufacturer. If so, these varying definitions for each manufacturer were not made explicit in their article [7.1]. Therefore, their crash analysis results could not be replicated.

7.2.4.1. Ambiguity in DPM Definition

For example, the numerator DPM in Equation (7.1) might be defined as "median disengagements per mile per vehicle per manufacturer." This definition would be feasible for Delphi, the only AV manufacturer that made their data available at the vehicle level for both disengagements and miles (see Table 7.2). However, the three other manufacturers did not release both vehicle-level disengagement and mileage data in their CA DMV disengagement reports (see Table 7.2 Therefore, DPM for those manufacturers might be defined as "median disengagements per mile per month per manufacturer." Or DPM might also be defined as "cumulative disengagements per cumulative miles per manufacturer" (i.e., as a mean rather than a median).

7.2.4.2. Ambiguity in DPA Definition

The denominator DPA in Equation (7.1) cannot be defined as a median DPA across vehicles. The reason is that accidents per vehicle cannot be matched to disengagements per vehicle because of the lack of vehicle IDs for accidents. DPA might therefore indicate a mean DPA per manufacturer (i.e., cumulative disengagements divided by cumulative accidents per manufacturer). In other words, a mean DPA can be easily estimated by dividing the cumulative (i.e., total) disengagements by the cumulative (i.e., total) accidents for each manufacturer because no vehicle-level data is then required. In addition, the total disengagements and total accidents are readily available in the CA DMV reports [7.2] and [7.3], respectively.

However, all AV manufacturers supply both mileage and disengagements per month (just not per vehicle per month). In addition, the AV manufacturers also provide the date of each crash in the crash reports [7.3]. Therefore, DPA could also be defined as a median DPA across months, using the accidents per month per manufacturer and the DPM per month per manufacturer. In other words, DPA could also be used to estimate a median DPA estimated from the median of the monthly DPA estimate per manufacturer.

7.2.4.3. Failure to Replicate Median APM per Manufacturer

Because of these ambiguities in the definitions of DPM and DPA, this review couldn't replicate Banerjee et al.'s (2018) [7.1] DPM data. Hence, Equation (7.1) could not replicate Banerjee et al.'s (2018) [7.1, Table VII] "3. Median APM" estimates in Table 7.1 for the four AV manufacturers. In short, a reanalysis was unable to replicate Banerjee et al.'s (2018) [7.1] main crash rate comparison of "3. Median APM" to the human-driven CV crash rate, shown in Table 7.1 as "6. Rel. to HAPM."

7.2.5. AV Crash Rates and IRRs Estimated Using Standard Methods

Banerjee et al.'s (2018) [7.1] cumulative data in Table 7.1 for crashes (1. Accidents) and mileage (2. Miles) for the four AV fleets were used to estimate crude AV crash IRRs as per the standard epidemiologic method (Chapter 1, Table 1.1) employed in the rest of this book. The AV crash IRRs estimated using standard methods were Waymo 11.8, Delphi 25.3, Nissan 89.5, and GM 698.9. These are shown in Column 8 in Table 7.1. The AV crash IRRs were substantially different from Banerjee et al.'s (2018) [7.1] "6. Rel. to HAPM" metric" (see Table 7.1, "9. % Diff. Vars. 6 and 8."). These are different metrics, so the fact that they are different in the empirical data should not be surprising.

Note that the "7. AV crash rate," based on a mean, is smaller than the "3. Median APM" metric for three of the manufacturers (Waymo, Nissan, and GM Cruise). Suppose the "3. Median APM" metric is related to an actual median estimate as the word "median" would suggest. If so, then the median is greater than the mean in the crash rate distribution that Banerjee et al. (2018) [7.1] used, whatever it was. That indicates that the crash rate distribution (whether estimated across vehicles or months) is "negatively skewed" or "skewed to the left." In other words, there is a long tail of low crash rates pulling the mean down more than the median. This description is plausible because it is well established that crashes are rare events, so crash rate distributions are non-normal. That is precisely why epidemiologic metrics do not assume normal distributions and assume Poisson rather than Gaussian distributions to compare exposed and unexposed groups and estimate confidence ranges (see Chapter 1, Section 1.2).

Nonetheless, the simple method of dividing the cumulative crashes by the cumulative miles is the standard method in vehicle safety for estimating crash rates, as shown in Chapters 2, 3, 4, 5 and 6. For example, Banerjee et al.'s (2018) [7.1] crash rate estimates compared to this standard crash rate method ranged from a 9% underestimate for Delphi to a substantial 533% overestimate for GM Cruise (Table 7.1, column 9).

Regardless of the differences in the two crash rate metrics, both the crude IRR estimates and the Rel. to HAPM crude estimates of Banerjee et al. (2018) [7.1]) in Table 7.1 are invalid estimates of the true AV crash IRR. Both crude estimates of the relative crash rate contain the same substantial bias issues documented in the other chapters in this book, particularly unrecorded crash selection bias and site selection bias. Therefore, neither the standard IRR estimates nor the Rel. to HAPM estimates in Table 7.1 should be cited as valid because they both contain substantial upward biases. Consequently, the standard IRR estimates and the Rel. to HAPM estimates of Banerjee et al. (2018) [7.1] in Table 7.1 are not further considered in this book. Some of the known biases in Banerjee et al.'s (2018) [7.1] analysis methods are described in the Sections below.

7.2.6. Unrecorded Crash Selection Bias

Banerjee et al. (2018) [7.1] correctly avoided heterogeneity between AV manufacturers by separately analyzing the data for each AV manufacturer. However, Banerjee et al.'s (2018) [7.1] analysis method contained a substantial upward unrecorded crash selection bias in their crash rate comparison of AVs (Median APM metric) to CVs (HAPM metric). Furthermore, this same unrecorded crash selection bias also upwardly biased the AV crash IRR estimates made in the reanalysis in the current study from Banerjee et al.'s (2018) [7.1] data (see Table 7.1, "8. AV crash IRR").

The unrecorded crash selection bias occurred because Banerjee et al. (2018) [7.1] compared all AV crashes (police-unrecorded and police-recorded) with 2015 U.S. national CV crashes (police-recorded crashes only). In other words, these investigators failed to recognize and adjust for the substantial number of CV crashes that are never reported to the police or, if reported, the police do not put a de-identified record into a publicly accessible database (see Appendix E, Table E.1).

Unrecorded crash selection bias in Banerjee et al.'s (2018) [7.1] analysis can be adjusted using several methods discussed in other sections of this book. For example (1) assumptions about the percentages of CV crashes that are not police recorded (Appendix E, Table E.1; Chapter 3, Section 3.5); (2) using only police-recorded AV crashes (see Chapter 2, Section 2.4.1, Table 2.5); (3) using trends (see Chapter 2, Section 2.4.2, Figure 2.2; Chapter 5, Section 5.4, Figure 5.2).

In sum, Banerjee et al.'s (2018) [7.1] unique metrics were intended to arrive at a more valid estimate of crash rates using medians rather than means. But Banerjee et al.'s (2018) [7.1] extremely high AV crash rate estimates relative to CVs were subject to unrecorded crash selection bias because all crashes were counted for the AVs but only police-recorded crashes for the CVs.

7.2.7. Site Selection Bias

Banerjee et al.'s (2018) [7.1] analysis also contained site selection bias because the test sites for the AV manufacturers had substantially different CV crash rates. For example, GM Cruise tests are conducted solely within San Francisco. That city has a considerably higher CV police-recorded crash rate than the Mountain View test site used by Waymo/Google (Appendix K, Table K.2) or the U.S. nationally (Appendix K, Table K.3). As another example, the Waymo/Google tests were conducted in Mountain View and vicinity, with a lower police-recorded CV crash rate than the U.S. nationally (Appendix K, Table K.1). As with the earlier studies reviewed in this book (Chapters 2, 3, 4, 5 and 6), Banerjee et al. (2018) [7.1] did not control for CV differences between the AV test site and the CV comparison site. Thus the AV crash rate at the AV test site was biased by differences between the CV crash rate at that site and the comparison CV crash rate at a different site. In other words, Banerjee et al. (2018) [7.1] used a biased CV comparison group to assess whether AV crash rates were different from CV crash rates. One way to address this issue is by standardization of the AV crashes by the CV crash rates at the AV test site, with the CV crash rates at the comparison site (see Chapter 3, Section 3.4.1 and Appendix H).

7.2.8. Conclusion to Analysis of the Banerjee et al. (2018) Study

As with the Dixit et al. (2016) [7.5] study in Chapter 4 and the Favarò et al. (2017) [7.4] study in Chapter 6, this review could not provide any final corrected, adjusted, and revised table in a reanalysis for the Banerjee et al. (2018) [7.1] data and methods. The reason is that the Banerjee et al. (2018) [7.1] study did not provide sufficient information or description of its estimation methods to allow straightforward replication of its unique AV crash rate metrics. Furthermore, it was infeasible to adjust for the substantial unrecorded crash selection bias identified in Section 7.2.6 and the site selection bias identified in Section 7.2.7 in this Chapter. Hence, Banerjee et al.'s (2018, Abstract) [7.1] conclusion that "...autonomous vehicles are 15 – 4000× worse than human drivers for accidents per cumulative mile driven" could not be verified as valid. Hence Banerjee et al.'s (2018) [7.1] AV crash rate ratios are not further presented or discussed in this book.

References

7.1. Banerjee, S.S., Jha, S., Cyriac, J., Kalbarczyk, Z.T. et al., Editors, "Hands Off the Wheel in Autonomous Vehicles?: A Systems Perspective on over a Million Miles of Field Data," Paper Presented at in *2018 48th Annual IEEE/IFIP International Conference on Dependable Systems and Networks (DSN)*, Luxembourg City, Luxembourg, June 25–28, 2018, http://ssbaner2.cs.illinois.edu/publications/dsn2018/Paper.pdf, accessed May 19, 2021.

7.2. CA DMV, "Disengagement Reports," 2015–2018, https://www.dmv.ca.gov/portal/vehicle-industry-services/autonomous-vehicles/disengagement-reports/. Note: Available upon request. Email to AVarchive@dmv.ca.gov to request a digital copy of an archived report. Requests must include the year and manufacturer.

7.3. CA DMV, "Report of Traffic Collision Involving an Autonomous Vehicle (OL 316)," California Department of Motor Vehicles, 2014–2019, https://www.dmv.ca.gov/portal/uploads/2021/05/OL316.pdf, accessed August 14, 2021.

7.4. Favarò, F.M., Nader, N., Eurich, S.O., Tripp, M. et al., "Examining Accident Reports Involving Autonomous Vehicles in California," *PLOS ONE* 12, no. 9 (2017): e0184952, doi:10.1371/journal.pone.0184952, http://journals.plos.org/plosone/article?id=10.1371/journal.pone.0184952, accessed May 19, 2021.

7.5. Dixit, V.V., Chand, S., and Nair, D.J., "Autonomous Vehicles: Disengagements, Accidents and Reaction Times," *PLOS ONE* 11, no. 12 (2016): e0168054, doi:10.1371/journal.pone.0168054, https://doi.org/10.1371/journal.pone.0168054, accessed May 19, 2021.

7.6. NHTSA, "2015 Motor Vehicle Crashes: Overview," Traffic Safety Facts: Research Note DOT HS 812 318," U.S. Department of Transportation, National Highway Traffic Safety Administration, Washington, DC, 2016, https://crashstats.nhtsa.dot.gov/Api/Public/ViewPublication/812318, accessed May 19, 2021.

7.7. Banerjee, S.S., Personal communication by Email, June 8, 2021.

7.8. https://gitlab.engr.illinois.edu/DEPEND/autonomouscars, accessed June 9, 2021.

7.9. https://gitlab.engr.illinois.edu/DEPEND/autonomouscars/-/blob/master/ReportParsers.py, accessed June 9, 2021.

7.10. https://gitlab.engr.illinois.edu/DEPEND/autonomouscars/-/blob/master/Visualizations.ipynb, accessed June 9, 2021.

8

Overall Summary of AV IRR Estimates and Issues

Contents

Note: Refer to Appendix B on page 149 for an explanation of the abbreviations and the definitions of unique terms and phrases.

This book critically analyzed six scientific studies [8.1, 8.2, 8.3, 8.4, 8.5, and 8.6], published from 2015 to 2018, of prototype AV crash rates and their comparison to CV crash rates. Each chapter first confirmed the original AV and CV crash and mileage data used in a given study from its cited sources. The original data and results of a study were presented at the start of each Chapter 2, 3, 4, 5, 6, and 7. Then, the original data were put into a standard epidemiologic design for estimating a relative rate ratio and its confidence limits (Chapter 1, Table 1.1). That ratio is formally termed an incidence rate ratio (IRR) in Epidemiology. An IRR was estimated for each study's original data where feasible and here termed the "original" IRR. In Epidemiology, this original IRR is called a "crude" estimate because it has not yet been adjusted for any biases. The epidemiologic design is beneficial because it is a standard design that shows how each study compared its AV crash rate estimate to its CV crash rate estimate. In addition, an epidemiologic design helps identify biases that cause deviations of the IRR estimate from the true IRR in the population under study.

8.1. Overview

Each chapter identified one or more of 11 issues with each study's original data, analyses, or results. These 11 issues were grouped into three major categories:

1. Systematic errors (minor and substantial)
2. Selection bias (in crashes, mileage, or sites)
3. Best practice violations (combining heterogeneous companies, crash severities, or CV sites)

Six studies crossed with 11 issues gave rise to a summary matrix of 66 cells (6 studies × 11 issues). A detailed matrix with links of each entry to its Chapter section is given in Table L.2 in Appendix L. This detailed matrix is then summarized in Table L.1 in Appendix L. In Tables L.1 and L.2, the rows labeled "Orig." are based on each study's original data and analyses.

Where possible, the identified systematic errors were then corrected, the identified selection biases adjusted, and the identified best practice violations eliminated. Those revisions resulted in a final crash, mileage, crash rate, and IRR estimate for all studies where it was possible. These revised estimates are in the rows labeled "Final" in Tables L.1 and L.2.

The L.1 and L.2 matrices can be described either by their rows (the six studies) or by their columns (the 11 issues). Section 8.2 provides an overall summary of the six studies (the rows in Tables L.1 and L.2). Section 8.3 gives a general overview of the 11 issues (the columns in Tables L.1 and L.2).

8.2. Overall Summary of AV IRR Estimates

This section summarizes the initial, partially-revised, and final AV total crash IRR estimates for the first five studies [8.1, 8.2, 8.3, 8.4 and 8.5], reviewed in Chapters 2, 3, 4, 5 and 6, respectively. Figure 8.1 summarizes the original and final IRR estimates for the first five studies reviewed and analyzed in Chapters 2, 3, 4, 5, and 6. The sixth study by Banerjee et al. (2018) [8.6] was not included for reasons given in Chapter 7, Section 7.2.6.

Figure 8.1 shows that the final IRR estimates (large green squares) could only be made for three of the six reviewed studies. That is, these studies contained enough information that their data and analysis could be corrected for known errors and adjusted for known biases. The studies are Schoettle and Sivak (2015) [8.1] in Chapter 2, Blanco et al. (2016) [8.2] in Chapter 3, and Teoh and Kidd (2017) [8.4] in Chapter 5. The original IRR estimates for these three studies are shown as the "x" symbols. These initial or "crude" estimates are based on the original crash and mileage data analyzed by each study. Visual inspection of Figure 8.1 shows that for the Schoettle and Sivak (2015) [8.1] study reviewed in Chapter 2, the final estimate was lower than the initial estimate. However, the final IRR estimate was higher

FIGURE 8.1 **Summary of the main AV total crash IRR estimates discussed in this book. Y-axis:** IRR estimate. **X-axis:** The first five studies reviewed and analyzed in this book (Chapters 2, 3, 4, 5, and 6). **Blue horizontal line:** The conventional vehicle benchmark (IRR = 1). **"x" symbols:** The five initial or "crude" IRR estimates based on each study's original data. **Open circles:** Three partially-revised IRR estimates with known data errors corrected and known biases removed, but with some known remaining biases. **Green squares:** Three final AV total crash IRR estimates in the three studies [8.1, 8.2, 8.4] for which all known errors could be corrected and all known biases could be adjusted (see Chapters 2, 3, and 5). The confidence intervals are the exact 95% mid-P estimates (see Section 1.2).

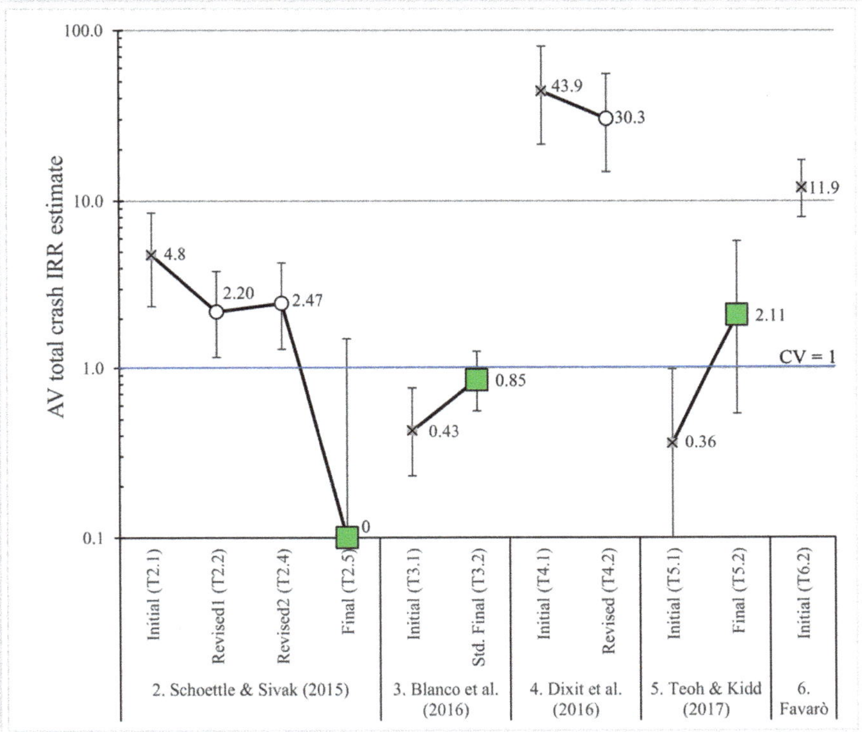

for the other two studies for which a final IRR estimate could be made: Blanco et al. (2016) [8.2] in Chapter 3 and Teoh and Kidd (2017) [8.4] in Chapter 5. Partially-revised IRR estimates (open circles) are also shown for the Schoettle and Sivak (2015) [8.1] study in Chapter 2 and the Dixit et al. (2016) [8.3] study in Chapter 4. The partly-revised estimates were lower than the original IRR estimates. Thus, only three studies had final IRR estimates (green squares) that can be regarded as relatively unbiased estimates of the true AV crash IRR.

For the Dixit et al. (2016) [8.3] study reviewed in Chapter 4, only the initial ("x" symbol) and partially-revised (open circles) IRR estimates are shown. For the Favarò et al. (2017) [8.5] study reviewed in Chapter 6, only the study's original IRR estimate is shown ("x" symbol). No final IRR estimate was possible in these two studies because error corrections and bias adjustments could not be completed with the amount of information available in the studies. Likewise, the study by Banerjee et al. (2018) [8.6] in Chapter 7 used a novel AV crash metric, median APM, that could not be replicated. However, this novel metric contained the same unrecorded crash selection bias and site selection bias as the original IRR estimates in the other studies and these biases could not be corrected from the information given by Banerjee et al. (2018) [8.6]. Therefore, Banerjee et al.'s (2018) [8.6] metric estimates are not included in Figure 8.1. The results for each of the six studies are briefly summarized below.

Chapter 2: Schoettle and Sivak (2015) [8.1]

Figure 8.1 (leftmost "x" symbol) plots the initial Schoettle and Sivak (2015) [8.1] IRR estimate of 4.8, which compared the relative AV total crash rate to the U.S. national total crash rate (Chapter 2, Table 2.1).

After correction and bias removal in Tables 2.2 and 2.3 and specific selections of the unknown percentages of unrecorded crashes for CVs, Figure 8.1 shows that the AV crash IRRs moved downwards to 2.20 and 2.47 (open circles). Thus, these revised IRR estimates are closer to an IRR of 1 (the CV crash rate at the blue horizontal line).

To eliminate unrecorded crash selection bias, only those AV crashes with police records were compared to the U.S. national police-recorded CV crash rates, eliminating unrecorded crash selection bias. Therefore, the final Google AV IRR estimate of 0 (Table 2.5) used only police-recorded AV and CV crashes, of which there were none among the Google AV crashes in the study period. Thus, the AV crash IRR for the Google AV data examined by Schoettle and Sivak (2015) [8.1] changed from causative (above 1) to protective (below 1) for the final AV IRR estimate after removal of as many of the identified issues as possible. The final IRR of 0 was assigned a value of 0.1 for plotting purposes on the log scale in Figure 8.1 (leftmost green square).

Chapter 3: Blanco et al. (2016) [8.2]

Using the original Blanco et al. (2016) [8.2] data in Table 3.1D in Section 3.2.1, a "protective" Google AV total crash IRR of 0.43 was estimated (Figure 8.1, "x" symbol) when compared to the SHRP2 CV crash rate.

Table 3.2 in Section 3.4.1 gives the revised estimated IRR after error correction and standardization to adjust site selection bias caused by the low CV PDO crash rate in Mountain View, CA. The Google AV final standardized crash IRR rose to 0.85 (green center square), near the CV crash rate of 1 (horizontal blue line). However, Figure 3.6 in Section 3.4.1 indicated that this increase in the standardized total IRR estimate was attributable solely to the increase in the standardized Major and Minor PDO crash IRR estimates.

Figure 3.6 in Section 3.4.1 demonstrated that the Google AV Severe crash IRR estimate slightly improved after standardization (from 0.64 to 0.57), so it could not have contributed to the increase in the standardized Google AV total crash IRR to 0.85 from the crude 0.43 estimate. Consequently, the 43% reduction in the Severe crash rate by the Google AV fleet has no known remaining biases, so it is a valid effect.

Chapter 4: Dixit et al. (2016) [8.3]

The initial Google AV IRR estimate of 43.9 is based on the initial data analyzed by Dixit et al. (2016) [8.3] in Table 4.1 in Section 4.2. (That initial estimate is plotted as the "x" symbol near the center of Figure 8.1.) This Google AV IRR estimate was artificially high in part because it had a substantial systematic error. The estimate used only police-recorded <u>injury</u> CV crashes rather than <u>all</u> police-recorded CV crashes for CA.

After correcting the error in Table 4.2 (Section 4.2.4), the Google AV IRR estimate decreased to 30.3 (open circle). However, this reduced Google AV IRR estimate was still upwardly biased because it contained unrecorded crash selection bias. The Dixit et al. (2016) [8.3] study compared all recorded plus unrecorded Google AV crashes to only police-recorded CA CV crashes. Dixit et al. (2016) [8.3] failed to adjust this upward unrecorded crash selection bias. Therefore, neither the initial nor revised Google AV IRR estimate should be cited as a valid estimate of the true Google AV crash IRR. In other words, neither estimate should be quoted as a valid estimate of the effect of Google AV exposure on crash rates.

Chapter 5: Teoh and Kidd (2017) [8.4]

Teoh and Kidd's (2017) [8.4] initial protective Google AV IRR estimate of 0.36 is shown as an "x" symbol in Figure 8.1. Table 5.1 in Section 5.1.1.5 indicates that the Google AV crash rate estimate was based on selected crashes that Teoh and Kidd (2017) [8.4] judged "police-reportable." The AV crash rate estimate was based on the Google prototype AV company fleet in Mountain View, CA, and vicinity and Austin, TX. Teoh and Kidd (2017) [8.4] selected the police-recorded CV crash rate in Mountain View for the comparison CV crash rate.

Teoh and Kidd's (2017) [8.4] study was the only study of the six that contained all 11 issues documented in Table L.1 in its original estimate. All 11 issues were a red "N," meaning "not OK," or "N/A," meaning "not addressed." In other words, 0% of Teoh and Kidd's (2017) [8.4] original data and analysis was "OK" (a green "Y").

Most of the 11 errors and biases were corrected and adjusted in the revised data and analyses. Table 5.2 in Section 5.1.3 estimated a final revised Google AV IRR of 2.11 (rightmost green square in Figure 8.1). The final Google AV IRR estimate was thus reversed in direction from the original protective IRR estimate of 0.36, the estimate based on Teoh and Kidd's (2017) [8.4] initial data. That is, the Google AV IRR estimate moved from protective (below the CV IRR of 1) to causal (above the CV IRR of 1) after correction and adjustment of the errors and biases in Teoh and Kidd's (2017) [8.4] analysis.

Chapter 6: Favarò et al. (2017) [8.5]

Favarò et al. (2017) [8.5] analyzed AV data for five "vehicle types" rather than for AV developers. The vehicle types were Nissan, Delphi/Audi, GM Cruise, a Google prototype, and a Google retrofitted Lexus.

Favarò et al.'s (2017) [8.5] initial AV crash rate estimate for each of the five "vehicle types" was substantially higher than the U.S. national CV crash rate estimate of 2 crashes per MVMT (Table 6.1, Section 6.1.1).

Hence, the initial pooled IRR estimate (Table 6.2, Section 6.1.2) based on the data was 11.9 (the "x" on the far right of Figure 8.1). An IRR of 11.9 is a substantial causative effect size. If free of bias, an AV crash IRR estimate that high would indicate that AVs cause substantially more crashes per MVMT than CVs.

However, Favarò et al.'s (2017) [8.5] initial comparison of these AV data to the U.S. national CV crash rate contained substantial unrecorded crash selection bias. The reason is that Favarò et al. (2017) [8.5] compared the rates of all AV crashes (police-recorded and not police-recorded) to the rates of only police-recorded CV crashes. Note: This unrecorded crash selection bias also substantially elevated the IRR estimates from Dixit et al. (2016) [8.3], reviewed in Chapter 4 (center symbols in Figure 8.1). It also substantially elevated the relative crash rate metric used by Banerjee et al. (2018) [8.6], described in Chapter 7 but not shown in Figure 8.1.

The Favarò et al. (2017) [8.5] study also had a heterogeneity best practice violation from combining crash and mileage data across "vehicle types" to form a Total AV estimate (Table 6.1, Chapter 6). Favarò et al. (2017) [8.5] had a second heterogeneity best practice violation because the study combined crash severities with substantially different AV crash rate estimates for each vehicle type (see Table 6.1, Section 6.1.1, Chapter 6). Like Dixit et al. (2016) [8.3], as described in Chapter 4, Favarò et al. (2017) [8.5] did not provide sufficient information about their data and methods to allow corrections or adjustments. As a result, the AV IRR estimate of 11.9 (Figure 8.1 and Table 6.2), based on Favarò et al.'s (2017) [8.5] initial data, is biased substantially high. Therefore, it should not be cited as a valid estimate of the effect of AV exposure on crash rates.

Chapter 7: Banerjee et al. (2018) [8.6]

The Banerjee et al. (2018) [8.6] study analyzed much of the same AV data as Favarò et al. (2017) [8.5]. The study also analyzed data from Google (later Waymo), Delphi, Nissan, and GM Cruise but analyzed it by AV developer rather than vehicle type. Also, their original AV relative rate estimates (see Chapter 7, Section 7.1) used a novel crash safety metric. Banerjee et al. (2018) [8.6] compared that metric for AV manufacturers to the preliminary 2015 U.S. national CV crash rate. The AV to CV crash rate ratio ranged from 20 to over 4,000 times higher, indicating substantial causal effects on AV crash rates.

However, all of Banerjee et al.'s (2018) [8.6] relative crash rate estimates had substantial unrecorded crash selection bias. The same unrecorded crash selection bias was in the Dixit et al. (2016) [8.3] study (Chapter 4 and center symbols in Figure 8.1) and the Favarò et al. (2017) [8.5] study (Chapter 6 and rightmost symbol in Figure 8.1). The reason is that all three studies compared the rates of all AV crashes (police-recorded and not police-recorded) to the rates of only police-recorded CV crashes. This unrecorded crash selection bias substantially increased the AV crash rate estimates relative to CVs, regardless of the AV crash rate metric that is used.

Banerjee et al.'s (2018) [8.6] relative crash rate estimates also had substantial site selection bias because they did not adjust for site differences in CV crash rates between the AV test site and the CV comparison site.

Banerjee et al. (2018) [8.6], like Favarò et al. (2017) [8.5], also committed heterogeneity best practice violations by combining data across AV manufacturers with substantially different crash rate estimates according to their metric. The study also likely committed heterogeneity best practice violations by combining data across crash severities without checking for homogeneity of its metrics. Like Dixit et al. (2016) [8.3] and Favarò et al. (2017) [8.5], this study did not provide sufficient information in its article about its data and methods to allow corrections or adjustments. Therefore, Banerjee et al.'s (2018) [8.6] AV to CV relative crash rate comparisons should not be cited as valid estimates of the effect of AV exposure on crash rates and are not considered further.

Final IRR Estimates with Fewest Remaining Issues

Three studies had the highest final test scores being "OK" on the 11 identified issues (Table L.1, second to the last column). In other words, the corrected and adjusted final IRRs of those studies eliminated almost all the known errors, biases, and best practice violation issues shown in Table L.1. Specifically, the final IRR estimate for the revised Schoettle and Sivak (2015) [8.1] (Chapter 2) data was 89% free of known issues, that from Blanco et al. (2016) [8.2] (Chapter 3) was 100% free, and that from Teoh and Kidd (2017) [8.4] (Chapter 5) was

FIGURE 8.2 Final AV total crash IRR estimates for the three studies in Figure 8.1 from which known errors could be corrected and known biases adjusted. Y-Axis: AV total crash IRR estimate. **Horizontal blue line:** Comparison CV IRR = 1. **Left green square:** In the final revised data of Schoettle and Sivak (2015) [8.1] (Table 2.5, Chapter 2), Google AVs lowered the total police-recorded crash rate to zero, estimating a zero Google AV IRR compared to CVs. (The zero IRR estimate is plotted at 0.1 because zero cannot be plotted on a log scale). **Middle green square:** In the final data of Blanco et al. (2016) [8.2] (Table 3.2, Chapter 3), standardized to remove site bias, the final Google AV IRR estimate was 0.85, about the same as SHRP2 CV total crash rate. **Right green square:** In the final revised data of Teoh and Kidd (2017) [8.4] (Table 5.2, Chapter 5), Google AVs slightly increased the total crash rate (IRR = 2.11) compared to CVs in Mountain View.

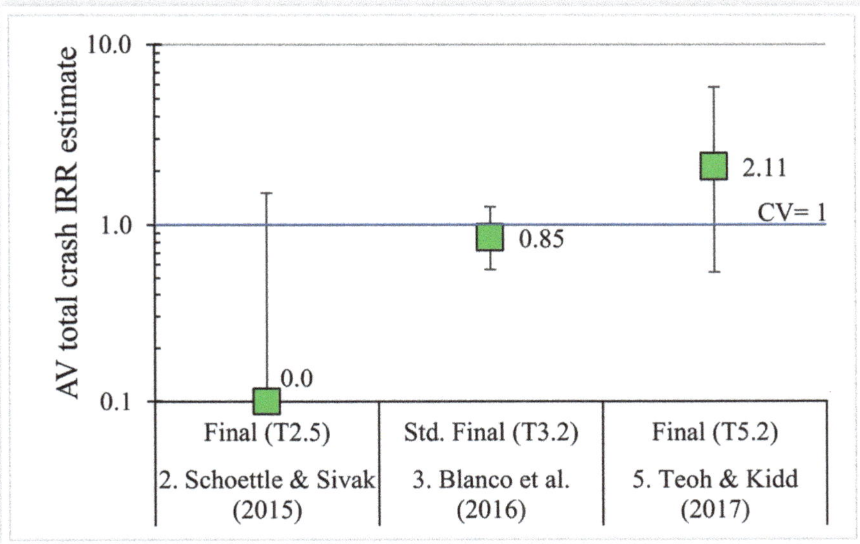

80% free of known issues. Thus, the three final revised and adjusted IRRs from these studies were judged to meet a 75% threshold for eliminating known errors, biases, and best practice violations (Table L.1, last column). Figure 8.2 replots (from Figure 8.1) the final AV total crash IRR estimates from these three studies, which provided enough data and information that most or all of the issues could be resolved by reanalyzing their data.

A test for heterogeneity using the Episheet (Rothman, 2015) [8.7] meta-analysis program found that the three IRR values were heterogeneous ($P = 0.049$). (The IRR of 0 was set to 0.001 for the heterogeneity test.) Heterogeneity indicates that the three studies have remaining unknown biases or errors. Otherwise, all the AV IRR estimates would be about the same effect size. The heterogeneity is self-evident in Figure 8.2 because one IRR estimate is below 1, another is near 1, and another is above 1 (where 1 is the CV IRR). This heterogeneity occurred even though all AV data analyzed in the final revised data for all three studies were from Google AVs.

The heterogeneity between these three AV total crash IRR estimates likely arose because each study used a different criterion for selecting Google AV crashes and a different CV reference dataset. For example, the final Google AV IRR of 0 for the reanalysis of the Schoettle and Sivak (2015) [8.1] data in Chapter 2 used only police-recorded AV crashes versus police-recorded U.S. national CV crashes (Section 2.3). The final reanalysis of Blanco et al.'s (2016) [8.2] data in Chapter 3 with an IRR estimate of 0.85 employed a standardized AV crash estimate in Mountain View and vicinity versus SHRP2 CVs at six sites (Section 3.4).

The final IRR estimate of 2.11 based on Teoh and Kidd's (2017) [8.4] corrected and revised data in Table 5.2 (Section 5.1.3, Chapter 5) used AV crashes judged by those investigators to be "police-reportable" versus Mountain View police-reported CV crashes.

Such heterogeneity makes it technically incorrect to arrive at an overall estimated IRR by pooling the IRR estimates across the three studies. So instead, it was here decided to pick the study with the fewest errors and biases remaining after the reanalysis and discuss its implications for AV safety.

The final revised IRR of 0.85 using the standardized Blanco et al. (2016) [8.2] data (Table 3.2, Section 3.4.1) had no remaining biases, errors, or best practice violations (Appendix L, Table L.1, Row 3 "Final"). It was the only study of the six studies free of unreported crash selection bias in its original data because it compared all crashes for fully instrumented AVs to all crashes for fully instrumented CVs. Therefore, Blanco et al. (2016) [8.2] was the only study of the six that did not have to make arbitrary assumptions about the percentage of unrecorded CV crashes in police records to adjust for unrecorded crash selection bias. Blanco et al. (2016) [8.2] also did not have to make decisions about which AV crashes would be "police-reportable" and which were not. For example, all the judgments about "police-reportable" AV crashes by Teoh and Kidd (2017) [8.4] were incorrect (Section 5.1.2.1). None of the Google AV crashes that Teoh and Kidd (2017) [8.4] identified as "police-reportable" through August 31, 2015, had an actual police record in the SWITRS database [8.8]. (Although, to be precise, the police may have written a PAR but then did not file a de-identified crash record in the publicly-accessible state crash database.)

In conclusion, after using standardization to adjust for site bias, the final Blanco et al. (2016) [8.2] IRR estimates (Table 3.2 and Figure 3.6) indicate that the Google AV fleet decreased the Severe crash rate by 45% versus SHRP2 CVs, with little change in the Major or Minor PDO crash rates. In other words, Google AVs reduced the Severe crash rate compared to CVs but had little effect on the Major and Minor PDO crash rates compared to CVs.

8.3. Overall Summary of Identified Issues

This section summarizes the identified systematic errors, biases, and best practice violations in estimating AV IRRs in the reviewed studies as totaled in the columns of Table L.1 in Appendix L. The links to the specific Section and Table are given in Table L.2 in Appendix L.

Issue 1. "Minor Systematic Error"

Four studies [8.1, 8.4, 8.5, 8.6] had "minor systematic errors." A "systematic" error means the error leads to results or conclusions that are systematically (as opposed to randomly) different from the truth. Three studies [8.1, 8.4, 8.5] had a "minor systematic error" in their original data compared to their sources. The fourth study [8.6] had a typo in one of its original calculated estimates. All known "minor systematic errors" were corrected in the final analyses and estimates.

Issue 2. "Substantial Systematic Error"

Three studies [8.3, 8.4, 8.6] had "substantial systematic errors" in their calculation methods or differences from their original data with their cited sources. In two studies [8.3, 8.4], the identified "substantial systematic errors" could be corrected in the final reanalysis. The third study [8.6] had a substantial systematic data anomaly in a novel AV crash metric

("Median APM") between the estimates in one of its Figures and the estimates in one of its Tables. The metric could not be replicated from the descriptions given in the study, so the metric values could not be corrected or adjusted in a reanalysis.

Issue 3. "AV Crash Selection Bias"

Two studies [8.1, 8.4] had an "AV crash selection bias" in their original tabulation of the AV crash count. This AV crash selection bias could be adjusted in one of the studies [8.1].

Issue 4. "AV MVMT Mileage Selection Bias"

Two studies [8.1, 8.4] had an "AV mileage selection bias" in their original tabulation of the AV MVMT. The bias was adjusted in the final reanalysis of both studies.

Issue 5. "Unrecorded CV Crash Selection Bias"

Five studies [8.1, 8.3, 8.4, 8.5, 8.6] had a substantial upward "Unrecorded CV Crash Selection Bias" in their original selection of CV crashes with a high percentage of unrecorded crashes. The bias was adjusted in the final reanalysis of two studies [8.1, 8.4]. This bias remained in the final analysis of three studies [8.3, 8.5, 8.6]. This bias was unrecognized and unaddressed in two studies [8.5, 8.6].

Overall, in Table L.1, it can be seen that Issue #5, "Unrecorded CV Crash Selection Bias," was tied with Issue #10, "Crash Severities Heterogeneity," as the second most common issue in the six studies reviewed in this book. This issue caused a substantial upward bias in the original AV crash IRR estimates in five of the six studies. The bias arose because different types of crashes were often selected for AVs and CVs, such as counting all crashes for AVs but counting only police-recorded crashes for CVs. Unrecorded crash selection bias arises from a failure to account for unrecorded CV crashes when comparing AV and CV crash rates.

The Schoettle and Sivak (2015) [8.1] study was the first to recognize unrecorded crash selection bias in comparing AV to CV data. It was also the first to attempt to adjust for it (Chapter 2, Section 2.1.2). Blanco et al. (2016) [8.2] (Chapter 3, Section 3.5) extensively investigated unrecorded crash selection bias and also attempted to adjust for it. However, their initial AV crash rates were compared to CV crash rates in instrumented vehicles where all crashes were recorded, as they were for AV crashes, so the bias did not exist for that comparison. Teoh and Kidd (2017) [8.4] also recognized the bias and developed a "police-reportable" class of AV crashes that was unfortunately not supported by actual police records (Chapter 5, Section 5.1.2.1). Unrecorded crash selection bias was not addressed by Dixit et al. (2016) [8.3] (Chapter 4, Section 4.2.2), Favarò et al. (2017) [8.5] (Chapter 6, Section 6.2.3), and Banerjee et al. (2018) [8.6] (Chapter 7, Section 7.2.6). Consequently, the original relative comparisons of the AV crash rate estimates to CV crash rates in these three studies (no matter what metric they each used for the AV to CV comparison) contained a substantial upward bias from unrecorded crash selection bias and should not be cited as valid.

A new method to avoid unrecorded crash selection bias using trend analysis was developed and illustrated in two studies (see Figure 2.2 and Figure 5.2).

Issue 6. "CV MVMT Mileage Selection Bias"

Two studies [8.4, 8.5] had a mileage selection bias in their original tabulation of the CV MVMT. The bias was adjusted in the final reanalysis of both studies.

Issue 7. "CV Crash Rate Differences at AV versus CV Sites," Site Selection Bias

All six studies [8.1, 8.2, 8.3, 8.4, 8.5, and 8.6] had a site selection bias in their original AV to CV crash rate comparison. Site selection bias was, therefore, the most common bias identified in this review. The bias arises from a CV crash rate difference in the AV site versus the CV site.

For example, the sites selected by Google for its AV testing (Mountain View and vicinity) had CV crash rates considerably different from the CV sites that every study chose for its AV-unexposed comparison, except study [8.4]. The CV comparison sites used in the six reviewed studies were as follows:

- [8.1]. 2013 U.S. national (Chapter 2, Tables 2.1, 2.2, 2.3, 2.4 and 2.5)
- [8.2]. Six SHRP2 CV sites (Chapter 3, Tables 3.1 and 3.2), 2013 U.S. national (Chapter 3, Section 3.5.1)
- [8.3]. 2014 California (Chapter 4, Tables 4.1 and 4.2)
- [8.4]. Mainly Mountain View, CA (Chapter 5, Tables 5.1 and 5.2)
- [8.5]. 2015 U.S. national (Chapter 6, Table 6.2)
- [8.6]. 2015 U.S. national (Chapter 7, Table 7.1, note f)

Because these CV sites had considerably different CV crash rates than the Mountain View site for the Google AV testing, the resulting site selection bias substantially biased the initial Google AV crash rate estimates, generally upward.

The bias was adjusted in the final reanalysis of the revised data of the Blanco et al. (2016) [8.2] study using a standardization method (Section 3.4 and Appendix H).

A new method to avoid site selection bias using trend analysis was also developed and illustrated using the Google AV crash data in two studies (see Figures 2.2 and 5.2).

Issue 8. "Non-CA AV Test Sites," Site Selection Bias

Two studies [8.1, 8.2] had an additional site selection bias because they included non-CA AV test sites in their analysis. States outside CA do not have any of the CA DMV AV crash and mileage reporting requirements. This bias was adjusted by removing the non-CA data in the final reanalysis of both studies.

Issue 9. "AV Companies Heterogeneity," Best Practice Violation

Two studies [8.1, 8.6] had an "AV Companies Heterogeneity" best practice violation by combining AV crash data across AV developers with substantially different (i.e., heterogeneous) crash rates. The bias was adjusted in the final reanalysis for one study [8.1].

Issue 10. "Crash Severity Heterogeneity," Best Practice Violation

Issue #10, "Crash Severity Heterogeneity," was tied with issue #5, "Unrecorded CV Crash Selection Bias," as the second most common issue in the six studies reviewed in this book. AV crash data were improperly combined across AV crashes with substantially different (i.e., heterogeneous) severities in five studies [8.1, 8.3, 8.4, 8.5, 8.6]. Study [8.2] recognized this as a potential issue and investigated it thoroughly. In addition, the bias was adjusted in the final reanalysis of the data from another study [8.1].

The issue arises because there are differences in the size of the AV effect on crash rates as a function of crash severity (see Study [8.2], Figure 3.6, Section 3.4.1, Chapter 3). Therefore it is necessary to stratify the data by crash severity so that the differential effect of AV exposure on crashes of different severities can be correctly assessed.

Issue 11. "CV Sites Heterogeneity," Best Practice Violation

This best practice violation of "CV Sites Heterogeneity" can occur when AV fleets have different ODDs, and these various ODD sites have heterogeneous CV crash rates. (This issue is related to the site selection bias issue #8, "Non-CA AV test sites.") Four studies [8.2, 8.4, 8.5, 8.6] had a best practice violation of heterogeneity in their selection of CV comparison sites. However, after reanalysis, only two studies [8.5, 8.6] still had an unaddressed CV site heterogeneity issue.

References

8.1. Schoettle, B. and Sivak, M., "A Preliminary Analysis of Real-World Crashes Involving Self-Driving Vehicles," Technical Report No. UMTRI-2015-34, The University of Michigan Transportation Research Institute, Ann Arbor, MI, 2015, http://www.umich.edu/~umtriswt/PDF/UMTRI-2015-34.pdf, accessed May 19, 2021.

8.2. Blanco, M., Atwood, J., Russell, S., Trimble, T. et al., "Automated Vehicle Crash Rate Comparison Using Naturalistic Data," Virginia Tech Transportation Institute, January 2016, https://vtechworks.lib.vt.edu/handle/10919/64420, accessed May 19, 2021.

8.3. Dixit, V.V., Chand, S., and Nair, D.J., "Autonomous Vehicles: Disengagements, Accidents and Reaction Times," *PLOS ONE* 11, no. 12 (2016): e0168054, doi:10.1371/journal.pone.0168054, https://doi.org/10.1371/journal.pone.0168054, accessed May 19, 2021.

8.4. Teoh, E.R. and Kidd, D.G., "Rage Against the Machine? Google's Self-Driving Cars versus Human Drivers," *Journal of Safety Research* 63 (2017): 57–60, https://www.researchgate.net/profile/Eric_Teoh/publication/317265345, accessed May 19, 2021.

8.5. Favarò, F.M., Nader, N., Eurich, S.O., Tripp, M. et al., "Examining Accident Reports Involving Autonomous Vehicles in California," *PLOS ONE* 12, no. 9 (2017): e0184952, doi:10.1371/journal.pone.0184952, http://journals.plos.org/plosone/article?id=10.1371/journal.pone.0184952, accessed May 19, 2021.

8.6. Banerjee, S.S., Jha, S., Cyriac, J., Kalbarczyk, Z.T. et al., Editors, "Hands Off the Wheel in Autonomous Vehicles?: A Systems Perspective on over a Million Miles of Field Data," Paper presented at in *2018 48th Annual IEEE/IFIP International Conference on Dependable Systems and Networks (DSN)*, Luxembourg City, Luxembourg, June 25–28, 2018, http://ssbaner2.cs.illinois.edu/publications/dsn2018/Article.pdf, accessed May 19, 2021.

8.7. Rothman, K.J., "Episheet: Spreadsheets for the Analysis of Epidemiologic Data," 2015, https://www.colleaga.org/tools/episheet, accessed May 19, 2021.

8.8. SWITRS, "California Highway Patrol Statewide Integrated Traffic Records System," https://iswitrs.chp.ca.gov/Reports/jsp/index.jsp, accessed May 19, 2021.

9

Overall Discussion

Contents

Note: Refer to Appendix B on page 149 for an explanation of the abbreviations and the definitions of unique terms and phrases.

This book critically reviewed the first six studies of AV crash safety published between 2015 and 2018 [9.1, 9.2, 9.3, 9.4, 9.5, 9.6]. The studies analyzed prototype AV test data on open roads in field studies with test drivers. The six studies analyzed AV test data collected for various periods from 2014 to 2018. The studies examined many of the same AV crashes. A single chapter (from 2 to 7) is devoted to each study.

The first section of each chapter shows the original data used in each study. Based on that original data, each chapter then estimates an AV crash rate and compares it to the CV crash rate selected by that study, using the standard epidemiologic method to estimate an IRR (Chapter 1). Each chapter next identifies the issues in each study, including systematic errors, biases, and best practice violations. Finally, as many of those issues as possible are addressed, and the data are reanalyzed. Final AV IRR estimates are then made on the reanalyzed data for the three studies with the most issues resolved.

9.1. Final AV IRR Estimates

Section 8.2 in Chapter 8 summarized the original AV IRR estimates for the six reviewed studies. Eleven issues concerning systematic errors, biases, and best practice violations were addressed in each study's data and analysis methods. Three studies had the fewest issues remaining in their final IRR estimates after correcting errors, adjusting for biases, and removing best practice violations (see Section 8.2 in Chapter 8 and Table L.1 in Appendix L). These three studies were (in order of fewest remaining issues):

1. The Blanco et al. (2016) [9.2] (Chapter 3) study, which was 100% free of the 11 issues in the final standardized data analysis, after starting at 67% free;
2. The Schoettle and Sivak (2015) [9.1] (Chapter 2) study, which was 89% free of the 11 issues in the final reanalysis of the revised data, after starting at 11% free;
3. The Teoh and Kidd (2017) [9.4] (Chapter 5) study, which was 80% free of the 11 issues in the final reanalysis of the revised data, after starting at 0% free.

The final AV IRR estimates of these three studies were heterogeneous: Figure 8.2 showed that one final IRR estimate was 0, another 0.85, and another 1.2. This heterogeneity in the IRR estimates indicates that one or more biases remained in the final estimates. There could not have been much remaining bias because all three studies attempted to estimate the same AV crash rate, using many of the same Google AV crashes, and had as many issues removed as possible. The one remaining bias was identified as site selection bias arising from crash rate differences in the CV comparison groups selected by the three studies.

The revised Blanco et al. (2016) [9.2] Google AV total crash IRR estimate of 0.85 (15% reduction in the overall crash rate) was selected as the most valid estimate of the Google AV population IRR estimate among the top three studies. First, the revised Blanco et al. (2016) [9.2] study data had the only AV crash IRR estimates that were 100% free of known issues after correcting errors, adjusting for biases, and removing best practice violations. Second, it was the only study of the six that stratified AV crashes according to crash severity: Severe, Major PDO, and Minor PDO. Third, the Blanco et al. (2016) [9.2] study provided the police-recorded CV data that could be used to remove the site selection bias by standardizing the Mountain View CV data.

Table 3.2A and Figure 3.5 (Section 3.4.1, Chapter 3) indicate that the SHRP2 comparison group in the Blanco et al. (2016) [9.2] study had 2.46 Severe CV crashes per MVMT. The Google AV data, after removal of site selection bias by standardization, had 1.41 Severe AV crashes per MVMT, a 43% reduction of the CV rate. This 43% reduction of the Severe crash IRR estimate after standardization was larger than the 36% reduction in the Severe crash IRR estimate in the Blanco et al. (2016) [9.2] original data (Figure 3.6). In other words, after standardization to adjust for site selection bias, the Google AV test fleet IRR estimate fell to 0.57, indicating a Severe crash rate reduction of 43% compared to CVs. Because all known biases had been removed, this Severe crash IRR reduction effect is an unbiased externally valid population effect. Therefore, one can make unbiased inferences from the Google AV Severe crash IRR estimate of 0.57 regarding the target population of Google AVs within their ODD.

However, after adjusting for site selection bias by standardization, the results of Google AV exposure for PDO crashes were not as beneficial as for Severe crashes. For the Google AV Major PDO IRR estimate, Figure 3.6 indicates an increase and reversal of direction from a <u>protective</u> IRR estimate (0.48) to a slightly <u>causative</u> IRR estimate (1.06). For the Google AV Minor PDO IRR estimate, Figure 3.6 indicates an increase from a <u>substantially</u> protective original IRR estimate (0.38) to a <u>slightly</u> protective IRR estimate (0.85) after standardization. For details, see Table 3.2 in Section 3.2.1 for the original Blanco et al. (2016) [9.2] data and Table 3.2 in Section 3.4.1 for the standardized data.

9.2. Final AV IRR Issues

Chapter 8 summarized eleven issues (systematic errors, biases, and best practice violations) in the six reviewed studies.

Section 8.3 in Chapter 8 found that Issue #7, Site Selection Bias, arising from CV crash rate differences at the AV versus CV sites, was the most common bias identified in this review. All six studies [9.1, 9.2, 9.3, 9.4, 9.5, 9.6] had Site Selection Bias in their original AV to CV crash rate comparison. Chapter 3 made a quantitative adjustment of the bias in the final reanalysis of the revised data in the Blanco et al. (2016) [9.2] study. The bias was also illustrated with trend analysis of the AV and CV crashes at the same site in Chapter 2 (Figure 2.2) and Chapter 5 (Figure 5.2).

Issue #5, "Unrecorded CV Crash Selection Bias," was tied with issue #10, "Crash Severity Heterogeneity," as the second most common issue in the six studies reviewed in this book. These issues were present in the original data and analyses in five studies [9.1, 9.3, 9.4, 9.5, 9.6]. Only the Blanco et al. (2016) [9.2] study did not have the issue in its original data and analyses.

9.3. New Findings

The main objective was to identify biases, errors, and best practice violations in the analysis of AV crash data so those issues can be reduced or eliminated in future investigations of AV crash data. However, the reanalysis to produce the revised and corrected results of the six studies led to five new findings:

1. The application of epidemiologic methods rather than conventional statistical significance tests makes it possible, even with the limited AV test data available, to make valid estimates of the relative risk of AVs versus CVs. Thus, using these methods, the field of AV real-world safety can move beyond the pessimistic outlook of Noy et al. (2018, Section 2.5) [9.7] that "the overall impact of AD on the frequency and severity of crashes is unknown, and there are insufficient data and understanding to permit valid estimates."
2. Epidemiologic methods applied to real-world AV test data reject the claim by Kalra and Paddock (2016) [9.8] that a valid assessment of AV crash risk requires hundreds of years and billions of miles of data collection.
3. The standardized Blanco et al. (2016) [9.2] data in Chapter 3 (Table 3.2 and Figure 3.6) indicate that prototype Google AVs with a safety driver from 2009 to October 31, 2015, reduced the Severe crash rate by 43% compared to SHRP2 NDS CVs, with smaller effects on Major and Minor PDO crash rates.
4. The reanalyzed Schoettle and Sivak (2015) [9.1] data in Chapter 2 (Table 2.5), using only police-recorded AV crashes, indicate that prototype Google AVs with a test driver from 2012 to September 30, 2015, reduced fatal, injury, PDO, and total police-recorded crashes to zero (100% reduction). This surprising lack of police records for AV crashes found here, even for known injury crashes, is partially consistent with the SHRP2 NDS result for CV crashes (Blanco et al., 2016) [9.2]. Table E.1 (last line) shows that the SHRP2 NDS dataset had 95% of CV crashes without a Police Accident Report after a careful search by Blanco et al. (2016) [9.2]. Many crashes with a PAR, particularly the Minor PDO crashes, are not filed by local police departments in the state police databases. Therefore, police-recorded crashes (in state and federal databases) have a smaller (and still unknown) percentage than the 5% of crashes with PARs. See Section 2.4.1 (Chapter 2) and

Appendix E for further discussion of this unreported and unrecorded crash selection bias issue. The low percentage of CV crashes without a PAR in the SHRP2 NDS suggests that the lack of police-recorded AV crashes is not likely a police bias against recording AV crashes compared to CV crashes. Although many AV crashes exceed state requirements for reporting the crash (see Appendix E) and are unrecorded, that is also the situation for a high percentage of CV crashes.

5. An upward trend was observed in the 2015 Google AV cumulative crash rate in Figure 2.2A (Section 2.4.2.3, Chapter 2). This same upward trend continued in 2016, as shown in Figure 5.2A (Section 5.4, Chapter 5). This AV upward trend was positively correlated with the upward trend in the CV cumulative crash rate (Figures 2.2B and 5.2B). However, the upward trend was negatively correlated with the downward trends in the AV cumulative disengagement rate for disengagements initiated by the test drivers (Figures 2.2C and 5.2C) and the ADS (Figure 5.2C).

These new findings warrant further research with the more recent CA DMV AV datasets available since 2018. Hopefully, this book will help future research studies of AV crash safety avoid the systematic errors, biases, and best practice violations identified in the six reviewed studies. If so, this book will have achieved its goal.

9.4. Limitations of the Reanalyses

1. The final AV crash IRR estimates for different crash severities in the standardized Blanco et al. (2016) [9.2] data (Table 3.2) include only prototype Google AV data. Therefore, those estimates may not generalize to the AV crash IRR estimates for other AV developers. In particular, the new finding of a 43% Severe crash rate reduction for prototype Google AVs from 2009 to October 31, 2015 (Table 3.2A) may or may not generalize to test results for other AV developers. Furthermore, these results may not generalize to subsequent test data from Waymo, spun off from Google to develop AVs and conduct AV tests.
2. The validity of the prototype AV IRR estimates in the reviewed studies was greatly limited by not having an adequately matched comparison group of CV crash rates. Five of the six studies contained issue #5, "Unrecorded CV Crash Selection Bias." Unfortunately, the issue was either unrecognized or incompletely adjusted by these studies. The study by Blanco et al. (2016) [9.2] was the only one of the six studies reviewed here that did not have this issue. That study used a comparison benchmark from a naturalistic driving study of heavily-instrumented CVs, which counted all crashes, as did the AVs.
3. However, the original data and analyses in all six studies, including the Blanco et al. (2016) [9.2] study, had site selection bias. In particular, the original data in all studies lacked a match in site conditions between the AVs and CVs. For example, except for the first few years of the Google AV test period from 2009 to September 30, 2015, the ODD of the Google AVs was geo-fenced to city streets with little or no highway and interstate driving. In contrast, the CA, national, and SHRP2 CV estimates included mileage from state roads, highways, and interstates. Additional crash and trip information provided by AV developer companies would help reduce this limitation. The Blanco et al. (2016) [9.2] study had its site selection bias removed by standardization. However, there was insufficient information in the other studies to standardize their data to remove their site selection bias.

4. It is unknown if AV developers design test regimens meant to put an AV through specific paces, in a certain subarea of a site or locale within their ODD. For example, for CVs, automotive manufacturers may run specific tests on the ride-and-handling loop at a proving ground, other tests on a cinder-block track, and yet other tests of tire friction on a high-speed roundabout. So when an AV company tests a prototype AV in San Francisco or Mountain View, or some other site, does the test driver have an assigned route that the AV must "drive" on any given day and an assigned test regimen that the AV must be put through? AV companies do not specify these details in the test procedures in their public reports. If there are assigned routes (say, to test left turns at busy intersections), the number of miles and the nature of those miles will not be representative or comparable to natural driving in a CV by a human driver. However, it is plausible that the AVs likely have a test route assigned to them by the AV developer for a given day. The vehicle occupants are probably given the freedom to redirect the route if they choose, at least within the limits of the geo-fenced autonomous driving area. More information provided by AV developers about AV test conditions on public roads will make it possible to improve matches of AV and CV site conditions (a valid match helps reduce site selection bias). Without such information from the AV companies, finding a valid comparison group of CV crashes and mileage is challenging.
5. A highly trained professional test driver was required in all AVs by California regulations [9.9] during the testing period of the six studies reviewed in this book. The test driver was there to "take over" driving control if the AV disengaged itself or the test driver chose to disengage the autonomous mode for whatever reason. There were also "planned" disengagements for testing purposes. Therefore, there are no open-road public crash data to estimate whether there would be a different AV crash rate with a non-trained ordinary driver as the "take-over" driver in an SAE Level 3 ADS. Likewise, at the time of the initial writing of this manuscript, there were not as yet any publicly available crash data for the driverless operation of an ADS-equipped vehicle as defined in Section 3.9 of revised SAE J3016_2021-04 (2021) [9.10].
6. There were new findings of an upward trend in Google AV cumulative crash rate and its positive correlation with the CV cumulative crash rate and its negative correlation with the test driver and ADS cumulative disengagement rates. However, these new findings were for only one AV developer (Google) at one site (Mountain View), during only two years (2015 and 2016). Therefore, these new findings may not generalize to other AV developers, other test sites, or other test periods.

9.5. Crash Rate Reduction: AVs versus Crash Avoidance Technologies

Yue et al. (2018, Table 4a) [9.11] reviewed the crash avoidance literature (both connected vehicle and driving assistance technologies). They found that crash avoidance technologies were effective in precrash scenarios in reducing police-recorded crash rates of human-driven light vehicles from a conservative 33% estimate to an aggressive 46% estimate. This effectiveness was despite consumer complaints (Jin et al., 2019) [9.12]. How do the crash rate reductions by AV technologies compare to the crash rate reductions in currently available crash avoidance technologies?

At first glance, the observed 43% reduction in Severe AV crash rates in the standardized Google AV data (Table 3.2A) is within the range of 33–46% police-recorded crash reduction rates in already commercially available crash avoidance technologies (Yue et al., 2018) [9.11]. That finding suggests that AVs, also equipped with crash avoidance technologies, do not provide a greater safety benefit than currently available crash avoidance technologies.

However, the answer is not as simple as it might appear. A deeper examination of this issue in Appendix J suggests that the Google AV Severe crash rate reduction of 43% (Table 3.2A) may not be readily comparable to crash avoidance technologies' crash rate reduction estimates. Therefore, a closer examination of the relative benefits of AVs, if any, to crash avoidance technologies in CVs should be undertaken in future research.

9.6. Did Lower AV Speeds Reduce the AV Severe Crash Rate?

The final standardized Google AV IRR estimates showed a 43% reduction in the AV Severe crash rate estimate compared to the SHRP2 CV Severe crash rate estimate (Figure 3.5 and Table 3.2). The speed hypothesis proposed here is that the slower AV driving speeds than human drivers caused the observed AV reduction in the Severe crash rate. It is well established that slower driving speeds lead to a lower crash rate, particularly for Severe crashes. For a review, see Aarts and Van Schagen (2006) [9.13].

It is also well established that the crashes that do occur are less serious at slower speeds (Aarts and Van Schagen, 2006) [9.13]). The standardized Google AV Major PDO crash rate estimate showed a 6% increase compared to the SHRP2 CV Major PDO crash rate estimate (Figure 3.5 and Table 3.2). Therefore, a corollary speed hypothesis is that the slower AV driving speed tends to convert Severe crashes into Major PDO crashes. If so, Severe crashes (which usually involve injuries) would decrease, but Major PDO crashes would increase. The percentage decline in Severe crashes would be larger than the percentage increase in Major PDO crashes because there are many more Major PDO crashes than Severe crashes. This effect is what the data indicate (Figure 3.5 and Table 3.2).

AV developers have not publicly released the speed limits they set in their AV ODDs. However, it is plausible that AV developers program the ADS to restrict AV driving speeds to the speed limit, or at most ten mph over the speed limit under certain driving conditions [9.14].

For CVs, Richard et al. (2020) [9.15] analyzed speeding behavior in the SHRP2 NDS. Chapter Six in that report [9.15] focused on the incidence of speeding behavior. The incidence of driving at least ten mph above the posted speed limit was high, with 99.8% of SHRP2 drivers having at least one of these speeding events within their trip sample. When the speeding data set was filtered to define speeding as driving at least 15 mph above the posted speed limit, 37% of drivers averaged about one event per 10 trips. Also, the incidence of speeding varied somewhat at the SHRP2 NDS sites in the six states. The average maximum speed for most speeding events ranged between 12 to 20 mph over the posted speed limit (Richard et al., 2020, p. 27) [9.15]. However, the average maximum speed in "riskier" speeders on 35 mph roadways exceeded the posted speed limit by an average of 65% (Richard et al., 2020, p. 75) [9.15]. The driver type identified as being in the "high speeder" group averaged at least one speeding event (more than 15 mph over the speed limit) per trip (Richard et al., 2020, p. 83) [9.15].

The association between speeding behavior and crash incidence and severity was not assessed in Richard et al. (2020) [9.15]. A test of the primary hypothesis is that if speeding more than 15 mph over the posted limit could somehow be eliminated, would there be about a 43% reduction in the Severe crash rate but not in the Major PDO crash rate? This hypothesis could be tested in future research using the SHRP2 data. The test should stratify the crash data by severity. Then, for each stratum, calculate the average speed of the crash-involved vehicles at the time of the crash. Then remove all crashes in which any crash-involved vehicle traveled at more than, say, 15 mph over the posted speed limit just before the crash. Then reestimate the crash rate at each crash Severity Level. If the Level 1 (Severe) crash rate declines, the primary hypothesis would be partially confirmed. If the Severe crash rate declines near 43%, the primary hypothesis would be fully confirmed.

9.7. Other Hypotheses to Explain the Upward Trend in AV Cumulative Crash Rate

Section 2.4.2.2 in Chapter 2 introduced four hypotheses to explain the upward trend in the 2015 Google AV cumulative crash rate observed in Figure 2.2A. Section 5.3 in Chapter 5 found that this same upward trend continued in 2016 in Figure 5.2A. Since the completion of those analyses, three additional hypotheses were developed, which are briefly considered here. More research is needed to test these hypotheses.

Hypothesis 5. Increased ODD Range

The Google AV cumulative crash rate rose in 2015 and 2016 because Google increased the number of roads within the range of its AV ODD in Mountain View and vicinity. Hence, the Google ADS neural networks would not have as much training on those roads, which would be expected to increase the Google AV crash rate.

However, Google never publicly stated that it had increased its ODD range. If anything, Google decreased its ODD range from its early initial testing period. For example, an AV crash occurred at 63 mph in March 2013 on Highway 680S in San Jose and another in March 2014 on Highway 101N near Belmont, CA (see Section 5.1.2.9). But there were no other crashes in San Jose or near Belmont after that. In addition, those two early crashes were on major highways under the jurisdiction of the CHP, and there were no reported crashes on major highways under the jurisdiction by the CHP after that. These results suggest that Google may have later restricted its ODD to city streets and not major highways under the jurisdiction of the CHP. (Note that "city streets" are defined as roads under the jurisdiction of local police departments, which can include state highways such as El Camino Real, which are no longer "major highways" but essentially city streets with intersections, stop lights, etc.) The Google AV ODD is also likely to still be mainly in Mountain View, with only limited AV mileage in nearby cities such as Palo Alto during the study period. Hence, an increased ODD range hypothesis seems implausible to explain the rising AV cumulative crash rate in Figures 2.2 and 5.2.

Hypothesis 6. Increased AVs

The rise in the AV cumulative crash rate in 2015 and 2016 at first appears consistent with the conjecture of Sivak and Schoettle (2015) [9.1] noted in Chapter 2 that, "During the transition period when conventional and self-driving vehicles would share the road, safety might actually worsen." Can Sivak and Schoettle's (2015) [9.1] conjecture explain why the Google AV cumulative crash rate rose during 2015 (Figure 2.2A) and 2016 (Figure 5.2)? As Google placed more AVs on public roads, the AV crash count would plausibly increase with a constant AV crash probability.

However, with more AVs on the road, AV mileage would also increase. Therefore, the AV crash rate would not increase simply from increasing AVs on the road during the potential changeover period with a mixed AV and CV fleet. Thus, the fact that the Google AV cumulative crash rate rose in 2015 (Figure 2.2A) and 2016 (Figure 5.2A) is inconsistent with Sivak and Schoettle's (2015) [9.1] conjecture. True, there might be more AV crashes because, with more AVs on the road, there would be more crashes between AVs and CVs just by probability. However, the AV crash rate would not change with the number of AVs on the road. Therefore, there has to be some other explanation than "increased AVs" for the upward trend in the Google AV cumulative crash rate in 2015 and 2016.

Hypothesis 7. CVs Are the Problem, not AVs

The analyses of the AV and CV cumulative crash rate data in Figures 2.2 and 5.2 indicated a worsening trend in the AV cumulative crash rate in 2015 and 2016. The analyses also found a strong association between AV and CV cumulative crash rates. As the CV cumulative crash rate rose, so did the AV cumulative crash rate. In other words, the worsening trend in the AV cumulative crash rate was predictable by the worsening trend in the CV cumulative crash rate. (Note: These findings were specific to Google AVs and Mountain View CVs and might not apply to other AV developers or other AV test sites.)

This rise in the AV cumulative crash rate was not explainable by Sivak and Schoettle's (2015) [9.1] conjecture about AVs and CVs sharing the road (see Hypothesis 6 above). However, there is a separate question about the crash rate between AVs if only AVs were on the road. The AV crash data in Figures 2.2 and 5.2 cannot answer this question because all reported AV crashes so far (except one) have been between AVs in autonomous mode and CVs.

Crashes between two or more CVs are relatively rare events. AVs are currently extremely rare on the road compared to CVs. Therefore, the probability of a crash between two AVs is far rarer than between two CVs, just from mathematics. The only known case of a collision between two AVs was on June 11, 2018, when a GM Cruise AV was rear-ended while operating in autonomous mode by a second GM Cruise AV operating in manual mode in a Minor PDO crash. If there are more AVs on the road, there is as yet no empirical data to answer the question of whether crashes between AVs would increase or not.

Hypothesis 7 suggests that "CVs are the problem, not AVs." Unfortunately, the data in Figure 2.2A show that not having a human driver in an AV did not reduce the upward trend in the AV cumulative crash rate trend, despite what AV developers and others had hoped (see Appendix A). But is the reason for the upward trend in the AV cumulative crash rate because of human CV drivers, not the AVs? AV developers implicitly promote this position because they almost always assign fault to the CV driver in a collision between an AV and a CV (see Appendix F). However, the close association between the upward trend in the CV cumulative crash rate and the upward trend in the AV cumulative crash rate (Figures 2.2 and 5.2) provides some evidence for Hypothesis 7. Assume the CV upward trend is the cause of the AV upward trend. An increased CV crash rate increases the probability of a crash with an AV, just like it increases the probability of a crash with another CV. So if an

AV crash would not have happened but for the CV driver, then eliminating CVs would theoretically stop the upward trend in the AV cumulative crash rate.

Assume that a site would fully convert to AVs so that AVs were the only vehicles on the road (AVs completely replace CVs at a given site). The question is whether the AV cumulative crash rate trend would continue upward, or would it turn downward over time? Hypothesis 7 postulates that the AV cumulative crash rate trend would turn downward compared to the AV cumulative crash rate upward trend evident in Figures 2.2 and 5.2. More specifically, Hypothesis 7 suggests that the AV cumulative crash rate trend might turn downward in proportion to the percentage of AVs in the overall fleet of vehicles at a given site.

However, there is no real-world empirical data on the crash rate between AVs operating in autonomous mode. Therefore, any further discussion of Hypothesis 7 is, at this point, pure speculation and conjecture.

References

9.1. Schoettle, B. and Sivak, M., "A Preliminary Analysis of Real-World Crashes Involving Self-Driving Vehicles," Technical Report No. UMTRI-2015-34, The University of Michigan Transportation Research Institute, Ann Arbor, MI, 2015, http://www.umich.edu/~umtriswt/PDF/UMTRI-2015-34.pdf, accessed May 19, 2021.

9.2. Blanco, M., Atwood, J., Russell, S., Trimble, T. et al., "Automated Vehicle Crash Rate Comparison Using Naturalistic Data," Virginia Tech Transportation Institute, January 2016, https://vtechworks.lib.vt.edu/handle/10919/64420, accessed May 19, 2021.

9.3. Dixit, V.V., Chand, S., and Nair, D.J., "Autonomous Vehicles: Disengagements, Accidents and Reaction Times," *PLOS ONE* 11, no. 12 (2016): e0168054, doi:10.1371/journal.pone.0168054, https://doi.org/10.1371/journal.pone.0168054, accessed May 19, 2021.

9.4. Favarò, F.M., Nader, N., Eurich, S.O., Tripp, M. et al., "Examining Accident Reports Involving Autonomous Vehicles in California," *PLOS ONE* 12, no. 9 (2017): e0184952, doi:10.1371/journal.pone.0184952, http://journals.plos.org/plosone/article?id=10.1371/journal.pone.0184952, accessed May 19, 2021.

9.5. Banerjee, S.S., Jha, S., Cyriac, J., Kalbarczyk, Z.T. et al., Editors, "Hands Off the Wheel in Autonomous Vehicles?: A Systems Perspective on over a Million Miles of Field Data," Paper Presented at *2018 48th Annual IEEE/IFIP International Conference on Dependable Systems and Networks (DSN)*, Luxembourg City, Luxembourg, June 25–28, 2018, https://ssbaner2.cs.illinois.edu/publications/dsn2018/Paper.pdf.

9.6. Teoh, E.R. and Kidd, D.G., "Rage Against the Machine? Google's Self-Driving Cars versus Human Drivers," *Journal of Safety Research* 63 (2017): 57–60, https://www.researchgate.net/profile/Eric_Teoh/publication/317265345, accessed May 19, 2021.

9.7. Noy, I.Y., Shinar, D., and Horrey, W.J., "Automated Driving: Safety Blind Spots," *Safety Science* 102 (2018): 68–78, https://doi.org/10.1016/j.ssci.2017.07.018, https://www.sciencedirect.com/science/article/pii/S0925753517304198?via%3Dihub, accessed May 19, 2021.

9.8. Kalra, N. and Paddock, S.M., "Driving to Safety: How Many Miles of Driving Would It Take to Demonstrate Autonomous Vehicle Reliability?," *Transportation Research Part A: Policy and Practice* 94 (2016): 182–193, doi:10.1016/j.tra.2016.09.010, https://www.researchgate.net/publication/308538761, accessed May 19, 2021.

9.9. CA DMV, "California Code of Regulations Title 13, Division 1, Chapter 1, Article 3.7—Testing of Autonomous Vehicles, §227.50," 2019, https://www.dmv.ca.gov/portal/uploads/2020/06/Adopted-Regulatory-Text-2019.pdf, accessed May 19, 2021.

9.10. SAE, "Taxonomy and Definitions for Terms Related to Driving Automation Systems for On-Road Motor Vehicles," SAE J3016_202104, SAE International, Warrendale, PA, 2021, https://www.sae.org/standards/content/j3016_202104/, accessed May 19, 2021.

9.11. Yue, L., Abdel-Aty, M., Wu, Y., and Wang, L., "Assessment of the Safety Benefits of Vehicles' Advanced Driver Assistance, Connectivity and Low Level Automation Systems," *Accident Analysis & Prevention* 117 (2018): 55–64, doi:10.1016/j.aap.2018.04.002, https://www.researchgate.net/publication/324531299, accessed May 19, 2021.

9.12. Jin, L., Tefft, B.C., and Horrey, W.J., Editors, "Mining Consumer Complaints to Identify Unsuccessful Interactions with Advanced Driver Assistance Systems," in *Proceedings of the 11th International Conference on Automotive User Interfaces and Interactive Vehicular Applications: Adjunct Proceedings*, Utrecht, the Netherlands, 71–75, 2019, ACM, https://dl.acm.org/citation.cfm?id=3351313 or https://doi.org/10.1145/3349263.3351313, accessed May 19, 2021.

9.13. Aarts, L. and Van Schagen, I., "Driving Speed and the Risk of Road Crashes: A Review," *Accident Analysis & Prevention* 38, no. 2 (2006): 215–224, https://saiv.espaceweb.usherbrooke.ca/References/010_2006_-DrivingSpeedAndTheRiskOfRoadCrashes_10p.pdf, accessed May 19, 2021.

9.14. Ingrassia, P., "Look, No Hands! Test Driving a Google Car," 2014, https://www.reuters.com/article/us-google-driverless/look-no-hands-test-driving-a-google-car-idUKKBN0GH02P20140817?edition-redirect=uk, accessed May 19, 2021. Note: The early Google prototype AV "Firefly" was limited to a 25 mph maximum speed. https://www.nbcnews.com/tech/tech-news/google-self-driving-car-gets-pulled-over-going-too-slowly-n462671, accessed July 12, 2021.

9.15. Richard, C.M., Lee, J., Brown, J.L., and Landgraf, A., "Analysis of SHRP2 Speeding Data," Department of Transportation, National Highway Traffic Safety Administration, Washington, DC, 2020, https://rosap.ntl.bts.gov/view/dot/44242, accessed July 12, 2021.

10

Overall Conclusions

Content

Note: Refer to Appendix B on page 149 for an explanation of the abbreviations and the definitions of unique terms and phrases.

Six scientific studies of AV safety published from 2015 to 2018 were based primarily on empirical AV crash and mileage data collected in prototype AV tests on open roads in CA. Eleven issues (systematic errors, biases, and best practice violations) were identified in the data and analyses of these studies. The predominant issues were:

1. Site Selection Bias (all six studies).
2. Unrecorded CV Crash Selection Bias (five of six studies).
3. Crash Severity Heterogeneity (five of six studies).

Only three studies (Schoettle and Sivak, 2015 [10.1]; Blanco et al., 2016 [10.2]; and Teoh and Kidd (2017) [10.3]) provided sufficient information that their original data could be replicated, and issues identified and removed in a reanalysis. However, the final AV crash IRR estimates across these three studies were heterogeneous after correction (one protective, one neutral, and one causative), likely because of differences in the selected CV comparison group.

The Blanco et al. (2016) [10.2] study, analyzed in Chapter 3, was the only study of the six that had no identifiable remaining issues after correction, adjustment, and reanalysis. It was selected as the study with the best estimate of the population AV IRR. That study compared the Google AV crash rates to crash rates in the SHRP2 naturalistic driving study that recorded all crashes. (AV developers must also record all AV crashes for tests conducted on public roads in CA [10.4].) After standardization of the Blanco et al. (2016) [10.2] data to remove site selection bias, the following four conclusions were reached (see Table 3.2 and Figure 3.6 in Section 3.4.1, Chapter 3). Compared to CVs, Google AVs are estimated to:

1. Reduce Severe crash rates by 43%.
2. Increase Major PDO crash rates by 6%.

3. Reduce Minor PDO crash rates by 15%.
4. Reduce total crash rates by 15%.

These conclusions should be tested in studies with the more recent AV data from 2019 onwards. Further studies should also be done on the AV crash rates from other AV developers. Hopefully, these further studies will avoid the errors, biases, and best practice violations identified in this book in these prior six studies. If so, the objective of this book will be met.

References

10.1. Schoettle, B. and Sivak, M., "A Preliminary Analysis of Real-World Crashes Involving Self-Driving Vehicles," Technical Report No. UMTRI-2015-34, The University of Michigan Transportation Research Institute, Ann Arbor, MI, 2015, http://www.umich.edu/~umtriswt/PDF/UMTRI-2015-34.pdf, accessed May 19, 2021.

10.2. Blanco, M., Atwood, J., Russell, S., Trimble, T. et al., "Automated Vehicle Crash Rate Comparison Using Naturalistic Data," Virginia Tech Transportation Institute, January 2016, https://vtechworks.lib.vt.edu/handle/10919/64420, accessed May 19, 2021.

10.3. Teoh, E.R. and Kidd, D.G., "Rage Against the Machine? Google's Self-Driving Cars versus Human Drivers," *Journal of Safety Research* 63 (2017): 57–60, https://www.researchgate.net/profile/Eric_Teoh/publication/317265345, accessed May 19, 2021.

10.4. CA DMV, California Code of Regulations Title 13, Division 1, Chapter 1, Article 3.7—Testing of Autonomous Vehicles, §227.50, 2019, https://www.dmv.ca.gov/portal/uploads/2020/06/Adopted-Regulatory-Text-2019.pdf, accessed May 19, 2021.

11

Specific Recommendations

Content

Note: Refer to Appendix B on page 149 for an explanation of the abbreviations and the definitions of unique terms and phrases.

1. Future empirical studies of AV safety should avoid the issues of substantial systematic errors, biases, and best practice violations identified in this critical analysis of six prior studies of AV crash rates compared to CV crash rates.
2. A more extensive study of AV crash and mileage data should be undertaken that follows the first recommendation. The new study should include all public AV data from all AV developers through the latest data available at the study time.
3. AV developers should provide more detailed crash and mileage information for past and future AV tests. The main advantage of releasing such information is that a more fully matched CV comparison group could be selected from police and naturalistic study databases. The information should include at least the following.

    a. For the AV involved in the crash: posted speed limit; crash Delta-V; the absolute speed at the time of the crash; maximum *g*-forces; direction of the impact; mileage when operating in manual mode; mileage when operating in autonomous mode; mileage on highways; mileage on city streets; traffic density at the time of the crash; road conditions; weather conditions; estimated repair cost of damage; whether airbags were deployed or not; whether the AV was towed; minimum 30-day follow-up for injury and fatality reports on the AV occupants; and vehicle identifier.

    b. For the CV(s) involved in the crash: absolute speed at the time of the crash; estimated repair cost of damage; whether airbags were deployed or not; whether the vehicle(s) were towed or not; minimum 30-day follow-up for injury and fatality reports on the CV occupant(s).

c. The AV mileage information in autonomous mode and manual mode for every city in the ODD. The goal is to improve site selection bias adjustment. The SHRP2 NDS dataset (e.g., as discussed by Blanco et al. (2016) [11.1]) offers an improved comparison CV group than do records of CV crashes in police databases. The NDS advantage is that NDS vehicles record all crashes, as do AVs. The national, state, or city crash data used in the other five studies reviewed in this book just used police-recorded crashes and either ignored or just estimated the percentage of crashes not recorded by the police. However, the SHRP2 sites were subject to site selection bias because the CV crash rates at the AV test sites did not match the CV crash rates at the SHRP2 sites. The release of this more detailed city-level mileage information will help reduce site selection bias in comparing AV to CV crash rates.

d. Although present in the early Google AV reports [11.2], the mileage in manual mode should be required in the CA disengagement reports [11.3], just like the mileage in AV mode must be reported to the CA DMV. Mileage in manual mode will make it unnecessary to standardize the AV data to remove site selection bias from differences in the CV crash rates at the AV site versus the CV comparison site. Also, crash and mileage data while AVs are in the manual mode need to be separated into use cases where crash avoidance technologies warn the driver or not. Previous monthly mileage in manual mode from all AV manufacturers for all prior months of the AV testing is also recommended for public release. Ideally, the CA DMV should require AV companies to release monthly mileage in manual mode in their yearly reports, just as they require monthly mileage in autonomous mode.

It is difficult to conceive how the public release of the information in a–d above would somehow place an AV company at a competitive disadvantage. It is not clear how by what means that information would somehow allow reverse engineering of AV designs. On the contrary, public release of such information would benefit both the general public and the AV developers. The public release would benefit AV companies by improving public trust. The public release would also help AV developers train the ADS neural networks to reduce crashes under the driving conditions of an AV crash from another company that their AVs have not encountered.

By providing more transparency and releasing the additional crash and mileage information suggested above, AV developers will make it easier to compare the AV crash data to NDS crashes or national CV crash databases in an unbiased manner.

4. Government AV policy changes.

   a. The current official policy position of the U.S. Department of Transportation (2018, p. 26) [11.4] is to rely on AV manufacturers to "demonstrate safety through voluntary safety self-assessments." Instead, U.S. government policy should require comprehensive de-identified public reports for all crashes and mileage on public roads for AVs in autonomous mode, during a transition, and manual mode.

   b. Governments should adopt the policy position advocated by Claybrook and Kildare (2018) [11.5] and Banks et al. (2019) [11.6]. Namely, governments should have greater oversight of AV testing and development.

c. Governments should require complete public reporting and transparency of all AV test results on U.S. public roads beyond the minimal reporting requirements now required in CA. However, such requirements were not in the proposed Federal AV rules (2018) [11.7], which mainly seek to reduce barriers to AV development, not to ensure safety; e.g., see Teoh (2019) [11.8].

d. The Federal government should adopt uniform police reporting requirements for AV crashes.

5. Future AV safety research should determine how many AV miles are required to demonstrate with 95% confidence that AVs reduce fatal crashes to zero compared to CVs.
6. Future AV crashes should classify injuries in AV crashes into the same injury classifications used in police records (severe, other visible, complaint of pain). The current study found that the Severe AV crash rate is reduced by 43% the Severe CV crash rate in the SHRP2 study. If the AV injuries were classified similarly to police records, then an AV to CV comparison can better be made to police records.
7. The hypothesis should be investigated that AVs reduce the Severe crash rate because they do not exceed the speed limit as frequently or as much as CVs.

References and Endnotes

11.1. Blanco, M., Atwood, J., Russell, S., Trimble, T. et al., "Automated Vehicle Crash Rate Comparison Using Naturalistic Data," Virginia Tech Transportation Institute, January 2016, https://vtechworks.lib.vt.edu/handle/10919/64420, accessed May 19, 2021.

11.2. Google Reports, "Google Self-Driving Car Project Monthly Reports," Mountain View, CA, May 2015–November 2016. Note: Monthly reports are no longer publicly available from Google. 19 monthly Google Reports from May 2015 to November 2016 are available at https://github.com/wgetsnaps/ca-dmv-autonomous-vehicle-accidents/tree/master/google-pdfs, accessed May 19, 2021.

11.3. Google, "Google Self-Driving Car Testing Report on Disengagements of Autonomous Mode December 2015," 2015, https://www.consumerwatchdog.org/resources/cadmvdisengagereport-dec.2015.pdf, accessed May 19, 2021. Note: Cumulative autonomous mileage from September 1, 2014, to November 30, 2015.

11.4. U.S. Department of Transportation, "Preparing for the Future of Transportation: Automated Vehicles 3.0," Washington, DC, 2018, https://www.transportation.gov/av/3, accessed May 19, 2021.

11.5. Claybrook, J. and Kildare, S., "Autonomous Vehicles: No Driver…No Regulation?," *Science* 361, no. 6397 (2018): 36–37, doi:10.1126/science.aau2715, https://science.sciencemag.org/content/sci/361/6397/36.full.pdf, accessed May 19, 2021.

11.6. Banks, V.A., Stanton, N.A., and Plant, K.L., "Who Is Responsible for Automated Driving? A Macro-Level Insight into Automated Driving in the United Kingdom Using the Risk Management Framework and Social Network Analysis," *Applied Ergonomics* 8 (2019): 102904, https://www.sciencedirect.com/science/article/abs/pii/S0003687019301358.

11.7. FMCSA, "Federal Motor Carrier Safety Regulations Which May Be Barrier to Safe Testing and Deployment of Automated Driving Systems-Equipped Commercial Motor Vehicles on Public Roads (FMCSA-2018-0037-0205)," 2018, https://www.regulations.gov/document?D=FMCSA-2018-0037-0001, accessed May 19, 2021.

11.8. Teoh, E., "Insurance Institute for Highway Safety—Comments," July 26, 2019, https://www.regulations.gov/document?D=FMCSA-2018-0037-0205, accessed May 19, 2021. Note: The attachment at the bottom of the document is the entire comment.

Appendix A
AV Hopes and Modeling Studies

CONTENTS

Note: Refer to Appendix B on page 149 for an explanation of the abbreviations and the definitions of unique terms and phrases.

A common assumption among AV advocates is that AVs offer a safety benefit. That assumption is typically based on hopes and modeling studies, or what is known as "face validity."

The "face validity" of the AV safety assumption was sufficiently strong that it was used to justify substantial R&D investments by many companies. AV field testing began in 2009, with AV crash and mileage data collected by many AV development companies since. This book critically reviewed the first six scientific studies, which evaluated the first empirical AV field data. These studies and this review attempted to test, with empirical data, the assumption that AVs offer a safety benefit over CVs. However, independently of the data, is the underlying assumption that AVs offer a safety benefit logically valid? This appendix helps answer that question.

A.1. AV Advocate Hopes Are Based on Reducing Human Error

Because AVs do not have a human driver when in autonomous mode, AV advocates hope that AVs will improve safety by eliminating human error. The following three examples are typical of the many statements that could be cited.

1. A U.S. Department of Transportation (2018, p. 3) [1] report states, "Of all serious motor vehicle crashes, 94 percent involve driver-related factors, such as impaired driving, distraction, and speeding or illegal maneuvers." The report then states about AVs, "Their potential to reduce deaths and injuries on the Nation's roadways cannot be overstated."

2. Waymo (2018, p. 3) [2] states, "Every year, 1.2 million lives are lost to traffic crashes around the world…A common element of these crashes is that 94% involve human error…We believe our technology could save thousands of lives now lost to traffic crashes every year."
3. GM (2018, p. 3) [3] states, "Human error is a major contributing factor in 94 percent of crashes," and "At General Motors, we envision a future with zero crashes."

These statements are carefully worded *not* to constitute a formal safety claim with legal and regulatory requirements and consequences. The statements are perhaps better characterized as "hopes" or "aspirations" that AVs will one day improve safety by replacing human drivers. Everyone certainly hopes that driving safety can be improved. It is, therefore, essential to examine the logical validity of statements that AVs will reduce crashes.

The above three AV safety statements are not written in a formal logical style. The statements are translated into a valid syllogism to reveal their implicit underlying logic.

A.2. AV Safety Hopes Are Based on an Implicit Syllogism

The underlying logic in the above three statements and their many variations regarding AV safety hopes and aspirations can be essentially reduced to a simple syllogism:

1. Major Premise: 94 percent of vehicle crash causes are human driver errors.
2. Minor Premise: AV crashes are not human driver errors.
3. Conclusion: Therefore, AV crashes are not 94 percent of vehicle crash causes.

The wording is somewhat clumsy to meet the syntax requirements of a formal syllogism validity checker [4]. However, the conclusion is valid, according to the logic checker in [4] as follows:

1. "All 9 are H."
2. "No A are H."
3. "Therefore, No A are 9."

The conclusion that "AV crashes are not 94 percent of vehicle crash causes" implies that AVs will reduce crashes by 94%. It is logical, as proven by the syllogism, and appears convincing to many (i.e., it has "face" validity). It is the basis for the "hope" of AV advocates in the previous section.

However, a critical analysis finds three errors in the assumptions of the syllogism: (1) Its Major Premise is misstated; (2) Its Conclusion is overgeneralized; (3) The syllogism contains an oversight.

A.2.1. Error 1: The Syllogism's Major Premise Is Misstated

The Major Premise is based on NHTSA's (2008, Table 8A) [5] analysis of on-scene investigations of 5,471 light vehicle tow-away crashes from July 3, 2005, to December 31, 2007. NHTSA classified 5,096 (93.1%) of the 5,471 tow-away crashes as having their "critical reasons for critical pre-crash event attributed to drivers." This 93.1% "critical reason" estimate attributed to drivers was updated to 94% in two later NHTSA reports (Singh, 2015 [5]; Singh, 2018 [5]). These reports dropped one of the crashes, so 5,470 crashes were analyzed. Singh (2018) [5] is identical to Singh (2015) [5], except Singh (2018) [5] adds a limitations section at the end.

However, the NHTSA (2008) [5] on-scene investigations could not establish that the "critical reasons" being "attributed to drivers" were *necessary* causes. A *necessary* cause requires evidence that the crash would not have occurred *but for* those causes, which an examination only of crash cases cannot provide. Crash investigations can only assess *possible* contributing risk factors. Outcomes, whether a disease, death, or crash, are caused by a "constellation of components that act in concert" (Rothman et al., 2008, p. 6) [6]. Indeed, it is well established that a crash has many interacting risk factors (Dunn et al., 2018) [7]. Noy et al. (2018, Section 2.2) [8] state that "some crashes that are coded as human error may involve other factors that cannot be readily overcome by technology" and quote several examples.

In other words, driver error is only one of many possible risk factors that contribute to a crash. Thus the NHTSA (2008) [5] report nowhere states that driver error is a sufficient or necessary cause in the crashes it examined. If it were such a cause, and humans could magically drive without error, then 94% of crashes would be eliminated. It must be proven that a crash would not have happened *but for* the driver error. Just estimating the prevalence of a driver error in crash cases, as employed in the NHTSA (2008) report [5], cannot itself prove causation. A formal scientific method must be used to prove causation.

One such formal scientific method used in this book is the Incidence Rate Ratio. It can prove a particular driver error is a cause by comparing the crash rate when the driver error is present to the crash rate when the driver error is not present (see Table 1.1). In other words, crash causation by a risk factor cannot be established solely by investigation of crash cases with the driver error present. Specifically, the prevalence (even if very high) of a driver error in a short period before a crash is insufficient to reach a valid conclusion about causation. Crash cases with a driver error being present among other risk factors also cannot be used to establish that the driver error was one of many risk factors that contributed to causing the crash (see Young, 2015, Section 1.3.1) [9]. To reach a valid conclusion about crash causation requires, in addition, for all biases to be removed in the comparison of the crash rate when exposed to the risk factor to the crash rate when not exposed to the risk factor (Table 1.1). Then and only then is it possible to establish a causal link of a risk factor to crashes. In short, the mere presence of a risk factor before a crash provides no evidence that the risk factor caused the crash. A crash rate with the risk factor must be compared to the crash rate without the risk factor. Another method to get a valid estimate of cause is the Risk Ratio Method, which compares the prevalence of a risk factor during a crash to its prevalence in baseline driving without a crash (Young, 2015, 2017) [9, 10]. See Young (2015) [9] and Young (2017) [10] for further discussion.

Epidemiology allows for a formal scientific determination of causation using these methods. Evidence for causation by a risk factor can be formally assessed using epidemiologic methods that compare the crash risk when the risk factor is present to the crash risk when the risk factor is not present (Rothman et al., 2008) [6]. (See Table 1.1 in Chapter 1 in the main body of this book for the design of an epidemiologic method that achieves this objective.)

A.2.2. Error 2: The Syllogism's Conclusion Is Overgeneralized

The Major Premise applies only to crashes in vehicles driven by humans, not to crashes in vehicles driven by an ADS (i.e., AVs). It is a tautology that if no human drove a vehicle, there could be no vehicle crash that involved a human driver error.

However, assume that AVs could somehow "fix" every human driving error so they would never be repeated in the ADS. Even if that were possible, the Conclusion of the syllogism is still overgeneralized. There will still be an unknown number of crashes that AVs themselves uniquely cause (see Noy et al., 2018, Section 2.3) [8]. The simple reason is that AVs make errors that human drivers do not. Neural networks can only recognize objects or events they have been "taught." If a neural network encounters an unusual object or event on which it has not been trained, the AV cannot recognize it and may crash if it is present or occurs in the vehicle trajectory. There have already been well-publicized fatal cases of an AV crashing into a pedestrian pushing a bicycle across the road and another into the side of a tractor-trailer straddling the road. Unlike neural networks, human drivers recognize and respond appropriately to unusual objects or events not previously encountered. Such "edge" cases are common in everyday driving because there are so many of them, even though any particular unknown object or event is rare. Koopman et al. (2019) [11] state, "...fixing edge cases as they arrive might not improve safety appreciably if the population of edge cases is substantial due to the 'heavy tail' distribution problem" (Koopman, 2018) [12]. In other words, training a neural network to recognize a novel object or event from an essentially infinite pool of novel objects and events does not materially improve the probability that a new rare object or event will be properly detected and safely responded to. Classifying novel objects and events by neural network processing of LIDAR and video input data will be worse than human perception for the foreseeable future. The human brain and human visual perception are far more complex than those untrained in the field of visual neuroscience realize. For such reasons, AVs using neural networks for object and event detection will continue to have crashes that human drivers will not for rare "Black Swan" objects or events.

A.2.3. Error 3: The Syllogism Contains an Oversight

The syllogism overlooks a new crash type introduced by AVs in mixed fleets: human drivers crashing into AVs. AVs drive differently than humans by design intent to avoid human driver errors. As a result, AVs drive in unexpected ways compared to human drivers (e.g., unusual slowing, stopping, and starting behaviors and extremely cautious driving behavior at intersections). An unexpected vehicle behavior increases the risk that a human-driven vehicle will crash into it (Dixit et al., 2016) [13].

In addition, even if all the vehicles on the road were AVs, it is plausible that there will still be AV-to-AV crashes because of software or hardware issues. There is yet no data on whether a crash of one AV could trigger other AV crashes. For example, a road departure crash of one AV at the head of a long train of AVs on a high-speed interstate could conceivably give rise to multiple AV crashes if the rearward AVs kept following the lead AV's road departure.

Because of these three errors, AV safety statements are questionable if they are based on an implicit logic such as the syllogism in Section A.2 (or some variant that is not logically valid).

However, there is an alternative to relying on hopes and aspirations based on questionable AV safety assumptions. Additional AV crash and mileage field data are needed that are appropriately analyzed with valid scientific methods to assess whether AVs truly improve safety or not. Six scientific studies have attempted to quantify with actual data whether AVs are progressing toward improved safety or away from it. Therefore, these scientific articles must be carefully examined, and the issues affecting the validity of their data, results, and conclusions should be identified. These include systematic errors, unrecorded crash selection bias, site selection bias, and best practice violations. This book aims to make these issues explicit so that they hopefully are not repeated in future scientific empirical studies of AV safety. Or, if they cannot be resolved, at least the limitation should be fully acknowledged.

A.3. Modeling Studies

Some modeling studies claim that a valid estimate of AV safety is impossible without hundreds of years of AV driving data. Kalra and Paddock (2016) [14] at the RAND Corporation used a "failure mode" [15] model with conventional statistical significance criteria. They estimated how many miles AV vehicles would need to drive to determine whether the AV "failure rate" is "significantly" lower than the human driver "failure rate." For example, to achieve "statistical significance" with 95% confidence and 80% power of a 20% fatality rate reduction over human drivers, the RAND model estimates 11 billion miles would be required (e.g., 518 years of testing of 100 AVs 24 hours a day at 25 mph). Based on that estimate and the hope that AVs will eventually improve safety, Kalra and Groves (2017) [16] recommend that open-road AV testing be continued, even without evidence that AVs reduce fatalities. Such a procedure would be unacceptable, for example, in the drug industry. Drug companies must present test trial results for review by experts at the U.S. Food and Drug Administration before any new drug is approved for public release [17]. For example, Phase III trial participants for a new drug must be assessed for disease prevention and side effects, with half receiving the medication and half receiving a placebo. The results must show that a new drug is safe and effective to specific criteria before receiving FDA approval.

However, the RAND "failure mode" model, although it is appropriate in some fields, has issues when applied to AV safety. Kalra and Paddock (2016) [14] and Kalra and Groves (2017) [16] rely on conventional statistical significance criteria to make decisions about AV safety benefits or harm from their model. However, for meaningful estimates, traditional statistics require large samples, such as the RAND estimate of 11 billion miles and hundreds of testing years. However, other analysis methods can estimate valid effect sizes from small samples using non-Gaussian distributions such as the Poisson distribution. In particular, the science of epidemiology was developed to enable rapid and valid estimates of disease and death risk based on small numbers of cases. Modern epidemiology categorically rejects the use of statistical "significance" criteria to make pass-fail decisions (Amrhein et al., 2019; Wasserstein and Lazar, 2016) [18, 19], such as the statistical significance criteria used in the RAND model. Epidemiologic analysis methods applied to AV safety (Young, 2018) [20] are thus preferable to the RAND "failure-mode" model and traditional statistical methods. The reason is that epidemiologic methods can reach valid conclusions based on few cases.

Appendix B
Abbreviations and Definitions

Note: Key terms are in bold when used within a definition.

ADS. An abbreviation for **automated driving system**.

APM. An abbreviation for "accidents per mile." Unique to Banerjee et al. (2018) [22]. See Sections 7.1 and 7.2.3, Chapter 7. The term has multiple meanings as used in [22] depending upon whether crash and mileage information is available or not from an **AV** manufacturer on a per-vehicle basis. So "accidents" can be cumulative crashes, median crashes per vehicle, or median crashes per month for a given **AV** manufacturer. "Mile" can be cumulative miles, median miles per vehicle, or median miles per month for a given **AV** manufacturer.

automated driving system.

> The hardware and software that are collectively capable of performing the entire **DDT** on a sustained basis, regardless of whether it is limited to a specific operational design domain (**ODD**); this term is used specifically to describe a Level 3, 4, or 5 **driving automation system**. (SAE J3016, 2021, Section 3.2, p. 6) [21]

autonomous. A deprecated term in the **SAE** J3016 Standard (2021) [21].

> This term has been used for a long time in the robotics and artificial intelligence research communities to signify systems that have the ability and authority to make decisions independently and self-sufficiently. Over time, this usage was casually broadened to not only encompass decision making, but to represent the entire system functionality, thereby becoming synonymous with automated. This usage obscures the question of whether a so-called '**autonomous vehicle**' depends on communication and/or cooperation with outside entities for important functionality (such as data acquisition and collection). Some **driving automation systems** may indeed be autonomous if they perform all of their functions independently and self-sufficiently, but if they depend on communication and/or cooperation with outside entities, they should be considered cooperative rather than autonomous. Some vernacular usages associate autonomous specifically with full driving automation (Level 5), while other usages apply it to all levels of driving automation, and some state legislation has defined it to correspond approximately to any **ADS** at or above Level 3 (or to any vehicle equipped with such an **ADS**). (SAE J3016, 2021, Section 7.1.1, p. 34) [21]

AV. An abbreviation for **autonomous vehicle** or **automated vehicle**.

autonomous mode. An **AV** being driven by the **ADS**.

> The status of vehicle operation where technology that is a combination of hardware and software, remote and/or on-board, performs the **dynamic driving task**, with or without a natural person actively supervising the **autonomous** technology's performance of the **dynamic driving task**. An **autonomous vehicle** is operating or driving in autonomous mode when it is operated or driven with the **autonomous** technology engaged. [60, §227.02(a)]

autonomous vehicle. An informal term used in this book and elsewhere to describe an **ADS**-equipped vehicle operating in **autonomous mode**. In other words, a vehicle with an **SAE Level 3**, 4, or 5 **driving automation system**. Also often informally termed **self-driving** vehicle or **automated vehicle**. All autonomous vehicles in this book were an **SAE Level 3 driving automation system**.

bench report. A method by which local police jurisdictions may write up a crash in their jurisdiction that has been reported. Bench reports may get tabulated and publicly released by the city hall. Or, the local police department may publicly release de-identified tabulated information from bench reports on their website. The local police department may or may not choose to write up a formal **Police Accident Report** for a crash with a bench report. See Appendix E.2.

best practice violation. An analysis, method, or procedure that does not follow the best practices in a given field. In an epidemiologic analysis, a best practice violation would be combining strata that are **heterogeneous** with one another in some effect measure. In automotive safety, a best practice violation would be to combine groups with **heterogeneous crash rates**. For example, summing crashes with different **Severity Levels** or combining data from companies with **heterogeneous crash rates**. Distinct from **bias**, **limitation**, and **systematic error**.

bias. In epidemiology, a

> systematic deviation of results or inferences from truth...An error in the conception and design of a study—or in the collection, analysis, interpretation, reporting, publication, or review of data—leading to results or conclusions that are systematically (as opposed to randomly) different from truth...The term bias does not necessarily carry an imputation of prejudice or any other subjective factor, such as the experimenter's desire for a particular outcome. This differs from conventional usage, in which bias refers to a partisan point of view—to prejudice or unfairness. (Porta, 2014, p. 21) [23]

See any epidemiology textbook for bias examples in the medical field or Young (2018) [20] for bias examples in the driving safety field. Bias from flaws in the study design, analysis, and interpretation of results can sometimes be removed by re-analyzing the data. Distinct from **best practice violation**, **limitation**, and **systematic error**.

CA. An abbreviation for California.

CHP. An abbreviation for California Highway Patrol.

CI. An abbreviation for **confidence interval**.

Conditional Driving Automation.

> The sustained and **ODD**-specific performance by an **ADS** of the entire **DDT** with the expectation that the **DDT** fallback-ready user is receptive to **ADS**-issued requests to intervene, as well as to **DDT** performance-relevant system failures in other vehicle systems, and will respond appropriately. (SAE J3016, 2021, Table 1, p. 26) [21]

confidence interval.

> The conventional form of an interval estimate in frequentist statistics. If the underlying statistical model is correct and there is no **bias**, a confidence interval derived from a valid analysis will, over unlimited repetitions of the study, contain the true parameter with a frequency no less than its confidence level (often 95% is the stated level, but other levels are also used). (Porta, 2014, p. 54) [23]

The confidence interval ranges between its lower limit and upper limit, abbreviated **LL** and **UL**, respectively.

confidence limits. "The upper and lower boundaries of the confidence interval" (Porta, 2014, p. 54) [23].

conventional mode. "The status of the vehicle when it is under the active physical control of a natural person sitting in the driver's seat operating or driving the vehicle with the **autonomous** technology disengaged" [60, §227.02(d)]. Same as **manual mode**.

conventional vehicle.

> "A vehicle designed to be operated by an in-vehicle driver during all or part of every trip" (SAE J3016, 2021, Section 3.32.1, p. 23) [21]. In this book, conventional vehicle refers more specifically to "A vehicle with no **driving automation system** features that is designed to be operated by an in-vehicle driver throughout all trips" (SAE J3016 2021, Section 3.32.1, Example 1, p. 23) [21].

crash rate. Number of crashes divided by **MVMT**. See **cumulative crash rate** and **total crash rate**.

crash selection bias. Bias in the estimated effect of an **AV** exposure on the crash **IRR** that arises from the procedures used to select crashes in the analysis. An example is **unrecorded crash selection bias**, from selecting all **AV** crashes (**police-recorded** and **police-unrecorded**) and then comparing that **AV crash rate** to the **CV crash rate** of only **police-recorded** crashes. See **selection bias**.

crash Severity Level. (capitalized) See **Severity Level**.

crash severity level. (not capitalized) See **severity level**.

cumulative crash rate. Cumulative crashes divided by cumulative **MVMT** up to a given date. See **total crash rate**.

cumulative disengagement rate. Cumulative **disengagements** of a particular type divided by cumulative **MVMT** up to a given date. See **disengagement** and **total disengagement rate**.

CV. An abbreviation for **conventional vehicle**.

DDT. An abbreviation for **dynamic driving task**.

Delta-V. The difference in velocity of two vehicles involved in a crash at the time of the crash.

disengagement. A **transition** from **automated mode** to **manual mode** in an **ADS**-equipped vehicle. Three broad types are commonly recognized: (1) A pre-planned disengagement during an **ADS** test, (2) an **ADS** technology failure, and (3) road situations in which a **test driver**/operator judges it is necessary to take manual control of the vehicle to operate safely.

DMV. An abbreviation for Department of Motor Vehicles.

driving automation system.

> The hardware and software that are collectively capable of performing part or all of the **DDT** on a sustained basis; this term is used generically to describe any system capable of Level 1 to 5 driving automation. Note: This term is generic for any **SAE** Level 1 to 5 system. The specific term for a Level 3 to 5 system is "**automated driving system (ADS)**." (SAE J3016, 2021, Section 3.6, p. 7) [21]

dynamic driving task. All of the real-time operational and tactical functions required to operate a vehicle in on-road traffic, excluding the strategic functions such as trip scheduling and selection of destinations and waypoints" (SAE J3016, 2021, Section 3.10, p. 9) [21]. "including without limitation: object and event detection, recognition, and classification; object and event response; maneuver planning; steering, turning, lane keeping, and lane changing, including providing the appropriate signal for the lane change or turn maneuver; and acceleration and deceleration" [60, §227.02(g)].

environment selection bias. Selecting an **AV ODD** with different environmental conditions (e.g., milder weather) than the **CV** comparison. See **selection bias, site selection bias, road selection bias**, and Appendix K.

FHWA. An abbreviation for Federal Highway Administration.

FMCSA. An abbreviation for Federal Motor Carrier Safety Administration.

GM. An abbreviation for General Motors.

GPS. An abbreviation for Global Positioning System.

HAPM. An abbreviation for Human **APM**. Unique to Banerjee et al. (2018) [22]. See Section 7.1, Chapter 7.

heterogeneity.

> (1) Differences in stratum-specific effect measures. When such measures are not equal it is said that the effect measure is heterogeneous or modified or varies across strata.
>
> (2) In a meta-analysis, the variability in the intervention effects being evaluated in the different studies. It may be a consequence of clinical diversity (sometimes called clinical heterogeneity) or of methodological diversity (methodological heterogeneity), or both, among the studies. It manifests in the observed intervention effects being more different from each other than one would expect due to random error (chance) alone. (Porta, 2014, p. 134) [23]

HPMS. An abbreviation for Highway Performance Monitoring System [24].

incidence rate.

> The rate at which new events occur in a population. The numerator is the number of new events that occur in a defined period or other physical span. Most often, the denominator is the person-time at risk of the event during this period; it may instead be in other units, such as passenger-miles. (Porta, 2014, p. 134) [23]

incidence rate ratio. "The **incidence rate** in the exposed group divided by the **incidence rate** in the unexposed group". (Porta, 2014, p. 145) [23]. The **AV** incidence rate ratio is estimated as the **AV crash rate** estimate divided by the **CV crash rate** estimate.

IRR. An abbreviation for **incidence rate ratio.**

Level 1 crash. A **Severe crash**. See **Severity Level.**

Level 2 crash. A **Major PDO** crash. See **Severity Level.**

Level 3 crash. A **Minor PDO** crash. See **Severity Level.**

limitation. Characteristics of design or methodology that affected or influenced the interpretation of the findings from the research. Distinct from **best practice violation**, **bias**, and **systematic error**.

LL. An abbreviation for the lower limit of the **confidence interval**.

Major PDO. A **Severity Level** 2 crash in a **naturalistic driving study** (see Appendix G). In **Police Accident Reports** and **police-recorded crashes** in crash databases, a "Major PDO" crash usually refers to a non-injury crash where the rough police estimate of the repair costs is equal to or above a criterion dollar amount. The criterion dollar amount can vary by police jurisdiction and calendar year. The definitions of a "Major PDO" crash overlap in the **SHRP2** study, **PAR**, and **police records** in that there is never an injury, which automatically places the crash into a "severe" category in all definitions. However, a "Major PDO" crash in **PARs** and **police records** can contain a tow-away or airbag deployment, which would place the crash into the **Severe Crash** category rather than "Major PDO" category in the **SHRP2** study. In this book, Major PDO will be capitalized.

manual mode. An **AV** being driven by a human.

Median APM. An abbreviation for median accidents per mile. A term unique to Banjaree et al. (2018) [22].

Median DPM. An abbreviation for median **disengagements** per mile. A term unique to Banjaree et al. (2018) [22].

mid-*P*. One of the exact methods of estimating a **confidence interval**. "The mid-P value is defined as half the conditional probability of the observed statistic plus the conditional probability of more extreme values, given the marginal totals" (Hwang and Wang, 2001) [25].

Minor PDO. A **Severity Level** 3 crash in a **naturalistic driving study** (see Appendix G). In **Police Accident Reports** and **police-recorded crashes** in crash databases, a "Minor PDO" crash usually refers to a non-injury crash where the rough police estimate of the repair costs is below a criterion dollar amount. The criterion dollar amount can vary by police jurisdiction and calendar year. The definitions of a "Minor PDO" crash overlap in the **SHRP2** study, **PAR**, and **police records**. In this book, Minor PDO will be capitalized.

MVMT. An abbreviation for millions of vehicle miles traveled.

naturalistic driving study. A study such as **SHRP2** with a fully-instrumented **CV** in which considerable crash data is stored and crashes are recorded on video.

NDS. An abbreviation for **naturalistic driving study**.

NHTSA. An abbreviation for National Highway Traffic Safety Administration.

non-PR crash. A known crash without a confirmed **PAR**. The crash may have been reported to the police, but the police did not write up a **PAR**. See **PR**.

ODD. An abbreviation for **operational design domain**.

operational design domain. "Operating conditions under which a given **driving automation system** or feature thereof is specifically designed to function, including, but not limited to, environmental, geographical, and time-of-day restrictions, and/or the requisite presence or absence of certain traffic or roadway characteristics" (SAE J3016, 2021, Section 3.21, p. 17) [21].

PAR. An abbreviation for **Police Accident Report**.

PDO. An abbreviation for Property Damage Only.

Police Accident Report. Summarizes information regarding the crash, usually containing facts related to the investigating officer's crash and opinions, including assignment of fault. It may or may not be generated by an investigating police officer who responds to a request for assistance at the scene of a car crash. It can also be generated by a later verbal report at the police department by a party or parties involved in a crash. A Police Accident Report may or may not be submitted by a local police department to become a record in a state or federal crash database. If that is done, the crash would also then be **police-recorded**. See Appendix E.

police record. De-identified crash record filed by police in a publically accessible database, usually at the state level. See **police-recorded crash.**

police-recorded crash. A crash with a **Police Accident Report** for which the state or local police are known to have filed a record in a publically accessible database, usually at the state level. Personally identifying information is removed from a **Police Accident Report** if the police chose to transfer the crash information to a crash record in a state database. See Appendix E.

police report. Informal term for **Police Accident Report, PAR**, or **PR.**

police-reportable crash. Unique to the Teoh and Kidd (2017) [26] study (see Chapter 5, Section 5.1.1.1). A crash that was judged in that study to be likely to have been reported to the police. A police-reportable crash may or may not be an actual **police-reported crash** or an actual police-recorded crash.

police-reported crash. A crash known to have been reported to the police, usually by a person involved in the crash or a witness to the crash. A police-reported crash may or may not have a **Police Accident Report** and may or may not be a **police-recorded crash.** See Appendix E.

police-unrecorded crash. A crash for which the state or local police have not filed a record in a publically accessible database, usually at the state level. A police-unrecorded crash may or may not have been reported to the police. If reported to the police, it may or may not have a police **bench report.** If it has a police **bench report**, it may or may not have a **Police Accident Report.** See Appendix E.

PR. An abbreviation for a **police-reported** crash as uniquely defined in the Blanco et al. (2016) [27, p. ix] study. That study used the PR abbreviation to mean "those crashes within the [**SHRP2**] database for which the research team has a confirmed **PAR.**" In other words, a PR crash is defined in [27] as a **police-reported crash** with a confirmed **Police Accident Report.** Note that in other studies, a **police-reported crash** can simply mean that the crash was reported to the police, and the police simply made a **bench report**, which the police did not turn into a formal **Police Accident Report.** Blanco et al. (2016) [27] did not investigate if the police also filed a given **SHRP2** PR crash in a state database to become a **police-recorded crash.**

PR crash selection bias. A **crash selection bias** that arises from comparing the **crash rate** for a group of **PR** crashes (crashes with a **PAR**) to the **crash rate** for a group of **police-recorded** crashes. See Appendix N.

road selection bias. Select an **AV** group that counts crashes and mileage on city streets only, but select a **CV** comparison group that counts crashes and mileage on all roads. See **environment selection bias, selection bias, site selection bias,** and Appendix K.

SAE. An abbreviation for Society of Automotive Engineers.

SAE Level 3. Conditional Driving Automation. All the studies reviewed in this book were of prototype **AVs** that could perform the entire **DDT** while engaged. However, all the prototype **AVs** in these studies had a "fallback-ready user, **test driver**," or "safety driver" who could assume control if the **ADS** automatically disengaged or if the safety driver chose to disengage the **autonomous** system. Hence, all prototype **AVs** in the studies reviewed here are classified during the **testing** phase as **Conditional Driving Automation (SAE Level 3)** as per the SAE J3016 Standard (2021, Section 5.4, p. 31) [21]. Note, however, that an **AV** developer may have the intention of releasing these prototype **AVs** (which were classified as SAE Level 3 during the **testing** phase) as **SAE** Level 4 or 5, which do not require a fallback-ready user.

selection bias. "**Bias** in the estimated association or effect of an exposure on an outcome that arises from the procedures used to select individuals into the study or the analysis" (Porta, 2014, p. 258) [23]. See **crash selection bias**, **site selection bias**, and **road selection bias**.

self-driving. A deprecated term in the **SAE** J3016 Standard (2021) [21].

> The meaning of this term can vary based on unstated assumptions about the meaning of driving and driver. It is variously used to refer to situations in which no driver is present, to situations in which no user is performing the **DDT**, and to situations in which a **driving automation system** is performing any part of the **DDT**. (SAE J3016, 2021, Section 7.1.3, p. 35) [21]

Severe crash. (capitalized) A **Severity Level 1** crash in a **naturalistic driving study (NDS)** (see Appendix G). See **Severity Level**. Compare **severe crash**.

severe crash. (not capitalized) In **Police Accident Reports** and **police-recorded** crashes in crash databases, a "severe crash" indicates a crash involving a severe injury to one or more parties involved in the crash. See **severity level**. Compare **Severe crash**.

Severity Level. (capitalized) Classifications of the seriousness of a crash in the **SHRP2 NDS**. Severity Level 1: **Severe crash**. Severity Level 2: **Major PDO** crash. Severity Level 3: **Minor PDO** crash. Severity Level 4: incidents not considered crashes (e.g., a tire strike on a curb). See Appendix G for detailed descriptions of Severity Levels. Severity Levels 1-4 should not be confused with the **SAE ADS** automation levels 1–5 (SAE J3016 standard, 2021) [21]. The Severity Levels are somewhat differently defined than the **severity levels** used in police reports and police records. Compare **severity level**.

severity level. (not capitalized) Classification of the seriousness of a crash in **Police Accident Reports** and **police-recorded crashes**. For example, Fatal, Injury (severe, observed, or verbal report only), **Major PDO**, and **Minor PDO**. The severity levels used by police may vary by police jurisdiction. The levels are somewhat differently defined than the **Severity Levels** used in the SHRP2 **NDS**. Compare **Severity Level**.

SHRP2. An abbreviation for the Second Strategic Highway Research Program (2006-2015). Used here to refer to the SHRP2 **naturalistic driving study** as described in Blanco et al. (2016) [27].

sIRR. An abbreviation for the **standardized incidence rate ratio**. See **standardization**.

site selection bias. Bias in the estimated effect of an **AV** exposure on the crash IRR that arises from the procedures used to select sites in the analysis. Example: A city **AV crash rate** is compared to the national **CV crash rate** without adjusting for differences in the city **CV crash rate** versus the national **CV crash rate**. See **environment selection bias**, **selection bias, road selection bias**, and Appendix K.

standardization. "A set of techniques, based on weighted averaging, used to remove as much as possible the effects of differences in age or other confounding variables in comparing two or more populations. The common method uses weighted averaging of rates specific for age, sex, or some other potential confounding variable(s) according to some specified distribution of these variables" (Porta, 2014, p. 268) [23].

standardized incidence rate ratio. An **incidence rate ratio** in which the numerator and denominator rates have been standardized (weighted) to the same (standard) population distribution. See **standardization**. Adapted from Porta (2014, p. 269) [23].

SWITRS. An abbreviation for Statewide Integrated Traffic Records System [28]. The police database for **police-recorded crashes** in **CA**.

systematic error. "...error that is consistently wrong in a particular direction; it often has a recognizable source (e.g., a faulty measuring instrument)" (Porta, 2014, p. 99) [23]. Systematic errors should be recognized and corrected before publication or after, if necessary. Distinct from **best practice violation**, **bias**, and **limitation**.

take over.

> A **DDT** fallback-ready user is expected to take over the **DDT** when a **DDT** performance-relevant system failure occurs or when the **driving automation system** is about to leave its **operational design domain** (**ODD**), then that user is expected to be receptive and able to resume **DDT** performance when alerted to the need to do so. (SAE J3016 standard, 2021, p. 24) [21]

take-over request. A request by the **ADS** for the person in the driver's seat to **take over** the vehicle's driving. In other words, a request by the **ADS** to **transition** from **autonomous mode** to **manual mode** driving.

test driver.

> A natural person seated in the driver's seat of an **autonomous** test vehicle, whether the vehicle is in **autonomous mode** or **conventional mode**, who possesses the proper class of license for the type of vehicle being driven or operated, and is capable of taking over active physical control of the vehicle at any time. [60, §227.02(c)]

testing. "The operation of an **autonomous vehicle** on public roads...for the purpose of assessing, demonstrating, and validating the **autonomous** technology's capabilities" [60, §227.02(o)].

TIMS. An abbreviation for Transportation Injury Mapping System [29].

total crash rate. Total crashes divided by total **MVMT** over a given period. See **cumulative crash rate**.

total disengagement rate. Total **disengagements** of a particular type divided by total **MVMT** over a given period. See **disengagement** and **cumulative disengagement rate**.

transition. A change from **autonomous mode** to **manual mode**. See **transition crashes**.

transition crashes.

> ...crashes that occurred when the vehicle was transitioning from **autonomous mode** and/or when the human driver of the automated vehicle (**AV**) was regaining manual control... Because **AV** drivers would not have been required to react to a **transition** (or make the decision to **transition**) in a "traditional" vehicle that was not equipped with **autonomous** capabilities, these crashes are considered to have occurred in **autonomous mode**. (Blanco et al., 2016, p. 8) [27]

UL. An abbreviation for the upper limit of the **confidence interval**.

UMTRI. An abbreviation for University of Michigan Transportation Research Institute.

unrecorded crash. A crash that has not been documented as a record in a police database at the state or federal level. A crash may be unrecorded for multiple reasons: (1) It may not have been reported to the police; (2) It may have been reported to the police and can be found in an informal **bench report**, but the police choose not to document it with a formal **Police Accident Report**; (3) It may have been reported to the police and be documented in a **bench report**, and **Police Accident Report**, but the police choose not to file a record in the police database at the state level. See Appendix E.

unrecorded crash selection bias. Selecting all **AV** crashes (**police-recorded** and **unrecorded**) and comparing them to only **police-recorded PR CV** crashes.

unreported crash. A crash that has not been reported to the police (but may have been reported to an insurance company). Or a crash that has not been reported to either the police or an insurance company. All unreported crashes are also **unrecorded crashes**. Compare **police-reported crash**, **police-recorded crash**, **PR crash**, and **police-unrecorded crash**. See Appendix E.

unreported crash selection bias. Selecting all **AV** crashes (**police-reported** and **police-unreported**) and comparing them to only **police-reported CV** crashes. See **police-reported crash**, **PR** crash, and **police-unreported crash**.

VMT. An abbreviation for vehicle miles traveled.

VTTI. An abbreviation for Virginia Tech Transportation Institute.

Appendix C
Transition Crashes: Descriptions and Comments

Contents

Note: Refer to Appendix B on page 149 for an explanation of the abbreviations and the definitions of unique terms and phrases.

Blanco et al. (2016, p. 8) [27] classified crashes during the transition from autonomous mode to manual mode as autonomous for reasons they state as follows:

> For purposes of this analysis, crashes that occurred when the vehicle was transitioning from autonomous mode and/or when the human driver of the automated vehicle (AV) was regaining manual control were considered autonomous mode. Because AV drivers would not have been required to react to a transition (or make the decision to transition) in a "traditional" vehicle that was not equipped with autonomous capabilities, these crashes are considered to have occurred in autonomous mode. (Blanco et al., 2016, p. 8) [27]

Here are the crash descriptions in the Google Reports [30] of two transition crashes:

> 2015 February: A Google Lexus model AV was travelling northbound on El Camino Real in autonomous mode when another vehicle travelling westbound on View Street failed to come to a stop at the stop sign at the intersection of El Camino and View Street. The other vehicle rolled through the stop sign and struck the right rear quarter panel and right rear wheel of the Google AV. Prior to the collision, the Google AV's autonomous technology began applying the brakes in response to its detection of the other vehicle's speed and trajectory. Just before the collision, the driver of the Google AV disengaged autonomous mode and took manual control of the vehicle in response to the application of the brakes by the Google AV's autonomous technology. The Google AV was in manual mode. No injuries were reported at the scene. The Google AV sustained some damage. (Google Reports, 2015, February) [30]

> August 20, 2015: A Google Lexus autonomous vehicle ("Google AV") operating in autonomous mode and traveling northbound on Shoreline Blvd. in Mountain View in lane two (the second of three lanes) was involved in an accident. As the Google AV approached the intersection of Shoreline Blvd. and High School way, a pedestrian began to cross the northbound lanes of Shoreline Blvd. in the crosswalk traveling westbound. The Google AV slowed to yield as it approached the crosswalk, and out of an abundance of caution the Google AV test driver disengaged the autonomous technology and took control of the vehicle. A vehicle in lane three to the immediate right of, and traveling in the same direction as the Google AV was already stopped and yielding the right of way to the pedestrian. A vehicle in the process of changing lanes from lane one into lane two and approaching from the rear struck the Google AV. The Google AV was travelling 5 mph at the time of impact, and braking to stop for the crosswalk. The other vehicle was travelling approximately 10 mph at the time of impact. The Google AV test driver reported minor back pain and was taken to a local hospital by Google employees, where he was evaluated and released by medical staff. The Google AV co-test driver did not report any injuries. The Google AV sustained minor damage to its rear left bumper. The other vehicle sustained moderate damage to its front end and was towed from the scene. The driver of the other vehicle did not report any injuries at the scene. (Google Reports, 2015, August) [30]

In both these transition crashes, the Google test driver intervened in the last few seconds before the crash and switched the vehicle from autonomous mode to manual mode. Such last-second transitions from autonomous mode to manual mode by test drivers' decisions are not uncommon.

These two transition crashes in the Google Reports [30] were counted as crashes in autonomous mode by three of the six studies reviewed in this book: Schoettle and Sivak (2015, p. 7) [31] (Chapter 2, Section 2.1.1), Blanco et al. (2016, p. 8) [27] (Chapter 3, Section 3.1.1), and Dixit et al. (2016, Table 1) [13] (Chapter 4, Section 4.1).

The CA DMV OL-316 AV crash reports [32] do not provide a checkbox for crashes during a transition. The only checkboxes provided in the OL-316 crash reports [32] are for "autonomous" and "manual." The AV companies invariably label transition crashes in the OL-316 AV crash report [32] checkboxes as "manual" crashes. However, whether the crashes are transition crashes can only be discerned by a careful reading of the details of the crash description text in the OL-316 AV crash reports [32].

These facts may be why three of the six reviewed studies did not mention transition crashes at all. These three studies classified crashes during a transition as crashes in manual mode (as they were scored on the OL-316 crash reports): Teoh and Kidd (2017, p. 58) [26] (Chapter 5, Section 5.1.2.2), Favarò et al. (2017) [33], and Banerjee et al. (2018) [22].

There are consequences of counting transition crashes as manual mode crashes and not as autonomous mode crashes. The test drivers can prevent a crash from being classified as autonomous if they voluntarily disengage the vehicle from the autonomous mode whenever they perceive a crash is imminent. There is nothing in the CA DMV AV rules that prevents a test driver from deliberately reducing the count of autonomous mode crashes by voluntary disengaging the autonomous mode before an imminent crash.

For some crashes, it is particularly self-evident that transition crashes should be counted as autonomous mode crashes. For example, in a crash on June 13, 2018, a Google test driver fell asleep during the autonomous mode, according to a contemporaneous article by an independent investigator (Efrati, 2018) [34]. The test driver disengaged the autonomous mode in his sleep (perhaps by unconsciously touching the steering wheel or pedals), and he "didn't respond to a bell sound notifying him that the car was now in manual mode, nor to a verbal audio alert from the car." The test driver reportedly woke up when "The vehicle hit and scraped against a median on its immediate left." The test driver falling asleep is not mentioned in the crash description in Google's report to the CA DMV database [32]. This tendency for drowsiness or even sleep in autonomous mode has also been noted in other AV crash reports. Counting this transition crash as a manual mode crash and not an autonomous mode crash downwardly biases the estimates of the AV crash rate and the AV crash IRR.

C.1. Do Test Drivers Cause Crashes When They Transition to Manual Mode?

John Krafcik, the Waymo CEO, has maintained[1] that test drivers' actions after voluntary disengagements are what "cause" a crash during a transition, not the ADS itself. If the driver had not voluntarily disengaged, Krafcik[1] maintains that the crash would not have happened based on a simulation of the crash. If such were the case, there would be little justification to classify a crash as "autonomous" during the short transition period from autonomous driving to manual driving just before a crash.

C.1.1. Waymo AV/Motorcycle Crash on October 19, 2018

The particular crash that Krafcik[1] was referring to, according to the Waymo OL-316 collision report,[2] was on October 19, 2018, on a three-lane roadway. The Waymo AV was going about 21 MPH in the center lane of three one-way lanes. A passenger vehicle in the left lane began to change lanes into the center lane. The Waymo test driver transitioned from autonomous mode to manual mode and began moving from the center lane to the right lane to avoid a potential collision. A motorcyclist previously traveling behind the Waymo AV in the center lane had already changed lanes from the center lane to the right lane and, going at about 28 MPH, attempted to pass the Waymo AV on its right. When the Waymo AV moved into the right lane, it sideswiped the motorcycle. The motorcyclist was injured and was taken for medical care. The police were called.

This crash is one of the rare AV crashes that ended up with a SWITRS [28] case record, #8727046. The SWITRS database was accessed through the TIMS [29] interface. A search by that case number for the Parties, Victims, and Collisions files shows that the police assigned fault to the Waymo test driver for an "unsafe lane change." The victim suffered "visible injury," which is not as serious as "severe" injury.

Waymo issued a press release which was covered in several newspaper accounts.[3,4,5,6,7] No record is mentioned of any lawsuits in the newspaper accounts.

---

[1] Krafcik, J., "The Very Human Challenge of Safe Driving," *Medium*, November 5, 2018, https://medium.com/waymo/the-very-human-challenge-of-safe-driving-58c4d2b4e8ee, accessed June 29, 2021.

[2] https://s3.documentcloud.org/documents/5029067/Waymo-Collision-Report-October-19-2018.pdf.

[3] https://www.businessinsider.com/waymo-self-driving-car-crash-sends-cyclist-to-hospital-driver-fault-2018-11.

[4] https://www.govtech.com/fs/in-waymo-self-driving-car-crash-human-driver-was-at-fault.html.

[5] https://www.washingtonpost.com/technology/2018/11/06/waymo-safety-driver-collides-with-motorcyclist-company-says-self-driving-minivan-would-have-done-better/.

[6] https://www.sfchronicle.com/business/article/Waymo-robot-car-injures-motorcyclist-but-13365135.php.

[7] https://thevintagent.com/2018/11/07/are-waymo-smarts-needed-to-protect-motorcyclists/.

There are several arguments against Krafcik's[1] position (that if the test driver had not intervened, no crash would have occurred).

1. It is not specified if the Waymo simulator was using the same software revision as was in the AV at the time of the crash.
2. The Waymo simulator has not had any study published to show that it can make valid predictions of real-world driving, as has been done for other driving simulators.[8]
3. The Waymo simulator's false positive and false negative prediction rates for real-world crashes have not been publicly released.
4. No interview was done with the motorcyclist to get another point of view.
5. Every crash reviewed in this book in which the OL-316 AV crash description [32] indicates a crash occurred during a transition, the test driver voluntarily took the vehicle out of autonomous mode. In other words, in the crash cases where the Waymo AV software or hardware failed, and the ADS requested the test driver to take over manual control, not a single crash was reported to have occurred during the transition. Suppose the test drivers were causing the crashes in the transition, as Krafcik maintains. Then why were there no crashes when the AV system itself made a take-over request to the test driver, who then switched the vehicle from autonomous to manual control? In short, all the crashes that occurred during a transition were only in those transitions where the test driver voluntarily initiated the transition to manual control, not when the ADS issued a take-over request to the test driver.

In other words, the usual transition crash is that a precipitating event occurred while the vehicle was in autonomous mode in the moments before a crash. Then the test driver perceived and believed an imminent crash was inevitable and quickly and voluntarily switched the vehicle out of autonomous mode just before the crash. In other words, the test driver perceived the AV had encountered an event or object and believed that the ADS-controlled vehicle behavior would not avoid the crash. In short, the test driver believed that leaving the vehicle in autonomous mode would have caused a crash.

Which scenario is more plausible? (1) An after-the-fact crash simulation indicating no crash would have happened if the test driver hadn't taken over? Or (2) a judgment by an experienced professional test driver in the real world that a crash was imminent? Suppose scenario (1) is more plausible. In that case, Krafcik's position[1] seems reasonable that a transition crash should be classified as a manual mode crash because it was the test driver who caused the crash by taking over manual control. If scenario (2) is more plausible, then assigning a transition crash to the vehicle's autonomous mode rather than its manual mode seems reasonable. Scenario (2) is consistent with the decision to assign transition crashes to autonomous mode made by Schoettle and Sivak (2015, p. 7) [31] (Chapter 2, Section 2.1.1), Blanco et al. (2016, p. 8) [27] (Chapter 3, Section 3.1.1), and Dixit et al. (2016, Table 1) [13] (Chapter 4, Section 4.1).

---

[8] Young, R.A., "Predicting Relative Crash Risk from the Attentional Effects of the Cognitive Demand of Visual-Manual Secondary Tasks," SAE Technical Paper 2017-01-1384, 2017, https://www.sae.org/publications/technical-papers/content/2017-01-1384/, https://doi.org/10.4271/2017-01-1384.

C.1.2. GM Cruise AV/Motorcycle Crash on December 7, 2017

With GM Cruise, there have been three motorcycle/AV crashes in San Francisco in 2017–2018 (December 7, 2017; August 7, 2018; and October 26, 2018). The December 7, 2017, GM Cruise crash with a motorcycle was similar to the Waymo crash with a motorcycle on October 19, 2018, except the GM Cruise AV was in autonomous mode. Of course, the GM Cruise ADS is entirely independent of the Waymo ADS. However, the GM Cruise example is a counterexample to the Waymo example. The reason is that the GM Cruise ADS did not prevent the sideswipe of the motorcycle on the right side, as Krafcik claims the Waymo ADS would have. In addition, a lawsuit was filed, so the viewpoint of the motorcyclist was available in the GM Cruise case. But the perspective of the motorcyclist in the Waymo case has not been made public.

The following crash description is paraphrased from the crash description in the CA DMV OL-316 December 7, 2017, GM Cruise crash report.

> The GM Cruise AV was in the center lane of three one-way lanes in heavy traffic, traveling at about 17 MPH. The GM Cruise ADS detected a gap between vehicles in the right lane and began moving the AV from the center lane to the right lane. But then the gap started to close, and the ADS stopped the AV from making the lane change and began re-centering the AV in the center lane. A motorcyclist driving between the center and right lanes at 17 MPH attempted to move into the center lane, and the two vehicles sideswiped. The motorcyclist was injured and was taken for medical care.

The following crash description is excerpted from the legal suit[9] filed by the motorcyclist:

> The Cruise AV, traveling directly in front of the motorcyclist, changed lanes to the left. Once the AV cleared the roadway, the motorcyclist proceeded to travel straight. However, at the same time, the AV suddenly veered back into the motorcyclist's lane, striking the motorcyclist and knocking him to the ground, and causing injuries.

So from the viewpoint of the motorcyclist, he was traveling in the same lane as the GM Cruise AV (contradicting the GM Cruise story that the motorcyclist was in between the center and right lanes). He saw the GM Cruise vehicle move to the left (GM Cruise description says the AV moved to the left to return to the middle lane after partially moving into the right lane). The motorcyclist did not say he had "attempted to move into the center lane." He says the GM Cruise vehicle veered into him (but GM Cruise says the motorcyclist attempted to move into the center lane). So the two stories are entirely different.

This GM Cruise case was the first legal case involving an autonomous vehicle and received extensive news coverage in the media, easily found in a Google search. The legal files can be viewed here.[10,11,12] A settlement was reached of an undisclosed amount, with the legal filings here.[13]

---

[9] https://storage.courtlistener.com/recap/gov.uscourts.cand.321643/gov.uscourts.cand.321643.1.0.pdf.
[10] https://www.law360.com/articles/1005140/gm-hit-with-first-known-suit-over-self-driving-car-crash.
[11] https://www.plainsite.org/dockets/39eyxv04d/california-northern-district-court/nilsson-v-general-motors-llc/.
[12] https://storage.courtlistener.com/recap/gov.uscourts.cand.321643/gov.uscourts.cand.321643.1.0.pdf.
[13] https://storage.courtlistener.com/recap/gov.uscourts.cand.321643/gov.uscourts.cand.321643.32.0.pdf.

C.2. AV Legal Issues Not in Scope

It is not within the scope of this book to go into legal issues concerning AVs. Some of these are discussed here.[14,15,16,17] The purpose of this book is solely to help elucidate how to identify and eliminate systematic errors, biases, and best practice violations in a scientific analysis of AV crash and mileage data. The GM case was brought up solely because it showed the two parties in a crash had opposite points of view. The alternative perspective was not brought out in a similar crash between a Waymo AV and a motorcyclist on October 19, 2018. The test driver had just transitioned the Waymo AV from autonomous mode to manual mode when the sideswipe of the motorcycle and the ride side of the Waymo AV occurred. This GM Cruise/motorcycle crash case was only brought up to show a counterexample to Krafcik's[1] claim that the Waymo/motorcycle sideswipe crash on the right side of the Waymo AV would not have happened if the test driver had left the vehicle in autonomous mode. The GM Cruise AV was in autonomous mode, yet there was still a sideswipe crash with the motorcyclist on the right side of the AV.

[14] https://medium.com/@marcdgrossman/the-new-norm-strict-liability-for-autonomous-automobile-accidents-38bbd0dc0a43.

[15] https://www.jstor.org/stable/26747441.

[16] https://www.icemiller.com/ice-on-fire-insights/publications/autonomous-vehicle-accidents-will-plaintiffs-crash/.

[17] Marchant, G. and Bazzi, R., "Autonomous Vehicles and Liability: What Will Juries Do," BUJ Sci. & Tech. L. 26:67, 2020, https://www.bu.edu/jostl/files/2020/04/3.-Marchant.pdf.

Appendix D
Should Crash Reports by AV Companies Be Accepted at Face Value?

Note: Refer to Appendix B on page 149 for an explanation of the abbreviations and the definitions of unique terms and phrases.

The AV data analyzed by five [13, 22, 26, 31, 33] of the six reviewed studies came solely from public crash descriptions written by the AV companies. The one exception was the study by Blanco et al. (2016) [27], sponsored by Google and given access to internal Google videos and instrumentation data from Google AV crashes. The public crash descriptions used by the studies were the Google Reports [30] and the OL-316 crash reports to the CA DMV [32]. But should these reports always be taken at face value? Or should there be an allowance for the possibility that AV companies may have minimized the severity of the crashes or left out or minimized other relevant information? For example, could the extent of damage to vehicles other than the AV have been underestimated? Or could injuries to other parties involved in the crash have been left out? What about injuries that first appeared or got worse more than ten days after the crash? (Note: Ten days after the crash is the required reporting deadline for OL-316 [32] crash reports.)

None of the six reviewed studies mentioned any attempt to search for contemporary newspaper accounts, Web postings of eyewitness statements, or photographs of the crash scene. Therefore, it must be concluded that the studies reviewed here, except Blanco et al. (2016) [27], accepted AV company crash reports at face value and used them as the sole basis for their data and analyses.

Such acceptance of AV company crash reports at face value might have underestimated crash severity relative to the crash severity assigned by police if they had been called to the scene. For example, would a PDO crash have met the $750 CA reporting criterion for a Major PDO crash if a police report had been made? (The criterion of $750 was the CA legal requirement for reporting a PDO crash during the collection of the CA test data.) However, might the crash falsely appear to be a Minor PDO crash from the slight damage described in a company's report? For example, when there are expensive sensors mounted on the AV exterior, how could any AV damage be under the $750 threshold?

As another example, one or more airbags may have deployed, in one or more vehicles, indicating a Severe crash by definition. However, airbag deployments were not mentioned by Google in at least one of its CA DMV OL-316 crash reports [32] (July 1, 2015; see Appendix E, Section E.4.1, and Appendix G).

Similarly, there were tow-aways in some AV crashes (July 1, 2015. and August 20, 2015; see Appendix E, Sections E.4.1 and E.4.2). Again, tow-aways were sometimes not mentioned in an AV company report [30, 32] (see Section E.4.1 in Appendix E). In addition, injuries to other parties or damage to other vehicles might not have been documented in the company report [30, 32] but might have been in a police report. Any such underestimates of AV crash severity would, in turn, have downwardly biased the AV crash counts, AV rates, and AV IRRs compared to police-recorded CV crashes.

Severe or even some Major PDO AV crashes have been considered newsworthy because of the keen public interest in (and concern about) AV safety. Although there is limited independent information on AV crashes, it is not difficult to find newspaper stories in a Web search for the more serious AV crashes. These often contain contemporaneous eyewitness accounts and even photographs of the crash scene.

As another example, in a Google AV transition side-impact, 30 mph, high Delta-V crash on September 23, 2016, Google's OL-316 AV crash report [32] says nothing about airbag deployment in the Google vehicle or a tow-away of either vehicle. However, a Web story on the same day by Hall (2016) [35] has a photograph of the crash scene showing both the front and side airbags were deployed in the Google vehicle. Furthermore, the photograph shows the van that crashed into the side of the Google vehicle being loaded onto a flatbed tow truck. Google's OL-316 AV crash report to the CA DMV [32] also does not indicate whether the local hospital treated the Google safety driver for injuries or not, just that he was "evaluated by medical staff and released."

In another example, a vehicle crashed broadside into a Google vehicle traveling 45 mph through an intersection on August 9, 2016, in Chandler, AZ, according to the Google Report [30]. However, no mention was made in the Google Reports [30] of airbag deployment or tow-away of either vehicle. Still, these were likely present because of the 45 mph speed.

Police records in the CA SWITRS [28] database, or crash reports in Mountain View or other city police public crash databases, can be searched for cases matched to AV crashes. Although matches are rarely found (see Section 3.4.1), such police records and Police Accident Reports, or even informal local police bench reports, where available, can provide an independent, objective description of AV crashes. Police descriptions of the crash are helpful because they provide independent information about damage to other vehicles in the crash or injury to parties other than the AV occupants. Unfortunately, the AV company reports [30, 32] seem to focus mainly on the damage to their own company AVs, with little information on the damage to other vehicles in the crash or injuries to other parties.

In summary, the possibility should be considered that crash reports from AV companies may understate:

A. Injuries
    1. The severity of injuries of AV occupants.
    2. The severity of injuries to occupants of other vehicles involved in the crash.
    3. The severity of injuries to pedestrians.
    4. Injuries that appear more than ten days after a crash, such as whiplash injuries.

B. Property Damage
    1. The extent of damage to AVs.
    2. The scope of damage to other vehicles involved in the crash.
    3. The extent of damage to property other than the vehicles, such as buildings, lamp posts, etc.

Appendix E
Why Unreported and Unrecorded CV Crashes?

Contents

Note: Refer to Appendix B on page 149 for an explanation of the abbreviations and the definitions of unique terms and phrases.

Why are so many CV crashes not reported to the police (i.e., police-unreported crashes)? Why do even more CV crashes not have a police record in a crash database (i.e., police-unrecorded crashes)? Many surveys have tried to answer that question for police-unreported crashes. However, surveys have limited ability to answer that question because they can only reflect the interviewed person's knowledge (and memory). Nonetheless, several survey studies have attempted to ascertain why respondents in a crash do not report the crash to the police. Or, if the respondent did report the crash to the police, why did the crash not end up being documented in a Police Accident Report or as a police-recorded crash in a police database at the state or federal level? The answers to such questions are essential to adjust for unrecorded crash selection bias. Also, do AVs have a lower proportion of police-reported and police-recorded crashes than CVs, even for a similar crash severity level?

E.1. Unreported CV Crashes

It is plausible that the number of Minor PDO crashes may be underestimated in surveys because such minor crashes may not be well recalled by drivers in a survey that occurs months after the crash.

Another reason for underestimates at all crash severity levels is that even if a driver recalls being in a crash, they may have deliberately chosen not to report it. For example, the Blincoe et al. (2015, p. 130) [36] survey asked respondents whether they recalled reporting a crash in the previous year to the police. Blincoe et al. (2015, p. 130) [36] give a variety of reasons why participants say they did not call the police for crashes they recalled being in:

"The most common was that the extent of vehicle damage or injury was not serious/severe enough to warrant reporting the crash, which together made up 56.7 percent of the responses for unreported injury crashes and 71.7 percent of the responses for PDO crashes."

Another reason for underreported crashes is if no other vehicles were involved in the crash. If other vehicles were involved, the probability of reporting increases: "The presence of a second driver, which automatically raises the issue of fault, increases the likelihood that crashes will be reported." (Blincoe et al., 2015) [36].

E.2. Non-PR and Unrecorded CV Crashes

Blincoe et al. (2015, p. 131) [36] note that "Reporting a crash to police does not assure that a police accident report will actually be filed. Individual police jurisdictions typically have reporting thresholds, especially for crashes that only involve property damage. Consumers may report crashes, but if police determine that the crashes do not meet the damage threshold, they may not file an accident report." A police-reported crash without a PAR is here termed a non-PR crash. In addition, just because the police write up a Police Accident Report does not mean they eventually file that as a record in the state police records database. Thus a PR crash may be an unrecorded crash, but all recorded crashes are PR crashes.

Complicating the issue is that Blincoe et al. (2015, p. 131) [36] note that "Reporting thresholds vary by State and sometimes by jurisdiction." Blincoe et al. (2015, p. 132) [36] add that "...seven States have no requirement for reporting crashes unless there is bodily injury, while one State requires reports only for tow-aways and one State requires all crashes to be reported." For a sample of six city or county police jurisdictions that are used for the National Accident Sampling System, Blincoe et al. (2015, Table 6-6) [36] show the total unreported crash rate varied from 4.8% for Cary, North Carolina, to 99.3% for Wake County, North Carolina, which are police jurisdictions in the same state.

In other cases, respondents in a telephone survey may say they correctly recall reporting a crash to the police. Still, the police may not document the crash in a formal Police Accident Report (PAR). If the police do create a PAR for a reported crash, then de-identified portions of that report might then get stored as records in crash databases at state levels, from which national crash databases often get their data. As a result, survey estimates of police-reported crashes at every crash severity level will always be higher than estimates of police-recorded crashes in public databases. Counts of police-recorded crashes are always fewer than counts of police-reported crashes, particularly for PDO crash severities. For example, the police are sure to record a reported fatal crash, but they may not record every reported injury crash. Likewise, the police are more likely to record a reported injury crash than a reported Major PDO crash. Also, a reported Major PDO crash is more likely to be recorded than a reported Minor PDO crash, which may not be recorded at all in many police jurisdictions.

The opposite relationship is true for unreported and unrecorded crashes. There are no unreported or unrecorded fatal crashes in the U.S. Unreported crashes at any severity level will necessarily be unrecorded. But given that a crash is reported, there will be fewer unrecorded crashes as crash severity increases. Note that unrecorded crashes necessarily include unreported crashes. However, unrecorded crashes can also be police-reported crashes for which local police jurisdictions only write a bench report. Such bench reports may get tabulated and publicly released by the city hall. Or, the local police department may publicly release on their website de-identified tabulated information from bench reports. In other words, the police do not write up a formal Police Accident Report for every reported crash. And if they do write up a Police Accident Report, they may still choose not to file de-identified portions as a formal record in the CA SWITRS [28] database or other state crash database in their jurisdiction is in other states.

Because of such validity issues and differences in practice, it is not surprising that there is a wide variation in the estimated percentages of unrecorded and unreported crashes, as shown in Table E.1.

TABLE E.1 Estimated percentages of unrecorded (**bold**) or unreported (*italic*) crashes: total, serious, and PDO.

Year of publication	Location	Period of data collection	Type of study	Total crashes	Total %	Serious %[d]	PDO %
	A. Telephone surveys						
1981[a]	U.S.	1981	Survey	579[b]	**45.6%**[a,c]	**24.4%**[c]	**48.9%**[c]
2011[e], 2015[f]	U.S.	12/2009-5/2010	Survey	2,252[g]	*29.3%*[h]	*15.4%*[h]	*35.6%*[h]
2015[i]	U.S.	12/2009-5/2010	Survey[j]	2,252[g]	**55.2%**[j,k]	**39.7%**[j,k,l]	**59.7%**[j,k]
	B. Naturalistic driving studies						
2006[m]	VA/DC[n]	2003-2006	100-Car	33[o,p]	**63.6%**	**0.0%**[q]	**75.0%**[r]
2006[m,s]	VA/DC[n]	2003-2006	100-Car	82[s]	**81.7%**[t]	**0.0%**[q]	N/A[r]
2016[u]	6 sites	2010-2013	SHRP2	912[o,p]	**95.0%**	**71.7%**[v]	**98.5%**[r]

Unrecorded crash percentage estimates are in **bold**. Unreported crash percentage estimates are in *italic*.

Notes:

[a] Greenblatt et al. (1981, p. 3-3) [37]. U.S. nationwide random telephone sample with 12,315 participating households. Verification of police-recorded crash used the National Accident Sampling System.

[b] There were 579 eligible crashes recalled by the respondents in the last four months during the telephone survey interview. Ineligible crashes were excluded, which "have no injury or damage and occur off the roadway (i.e., in parking lots, driveways, fields, etc.)" (Greenblatt, 1981, p. 4-19, p. 5-1 footnote) [37].

[c] Percentage of unrecorded crashes. These include: (1) crashes never reported to the police, plus (2) crashes reported to the police but for which the police chose not to file a formal police report. There were 78 serious (injury) crashes in the survey, of which 19 (24.4%) were unrecorded (Greenblatt, 1981, Table 5-2 bottom) [37]. The PDO crashes were estimated by subtracting injury crashes from the total crashes. There were 501 PDO crashes, of which 245 (48.9%) were unrecorded. However, the 24.4% unrecorded injury figure is an incorrect percentage for crashes because it represents injuries at the victim level, not the crash level. See Section E.2.1.2.

[d] "Serious" is defined as "injury" for surveys and Severity Level 1 for NDSs. Severity Level 1 crashes include injury or airbag deployment, tow-away, or high |g| forces even without injury (see Appendix G).

[e] M. Davis & Co. (NHTSA, 2011) [38], quoted by Blincoe et al. (2015, p. 130) [36].

[f] M. Davis & Co. (2015, Table 3.1.1a) [39]. A nationwide sample of 33,530 U.S. households.

[g] Total crashes at any prior time, as recalled by the respondent during a telephone survey interview.

[h] Unreported crashes as a percentage of total crashes.

[i] Blincoe et al. (2015, p. 13) [36].

[j] Revised. Based on survey data in notes e and f but adds estimated crashes reported but not recorded by the police.

[k] Estimated percentages of unrecorded crashes (unreported crashes, plus reported but unrecorded crashes).

[l] 39.7% = correct estimated percentage of unrecorded injury crashes in Blincoe et al. (2015) [36], not 24.3%.

[m] Dingus et al. (2006) [40].

[n] Northern Virginia and Washington, DC metro area.

[o] Total crashes with a fatality, injury, or PDO. Excludes Severity Level 4 crashes with low-|g| physical contacts (e.g., tire strikes) with no property damage (see Appendix G).

[p] Excludes crashes with incomplete data because of an occurrence during the original computer 90-second startup period, camera failure, etc.

[q] The definition of Severity Level 1 crashes changed from the 100-Car study [40] to the SHRP2 study results as presented in Blanco et al. (2016) [27] (see Appendix G). Crashes in the 100-Car study had to be a PR crash, or they were not classified as Severity Level 1. Severity Level 1 crashes in the SHRP2 study [27] did not have to be a PR crash. This change in definition explains the discrepancy in unreported injury crashes of 0% (note q) in the 100-Car study [40] versus 71.7% (note v) in the SHRP2 study [27].

[r] Severity Level 2 (Major PDO) and Severity Level 3 (Minor PDO) non-PR crash percentages.

[s] These 82 crashes incorrectly include 49 Severity Level 4 tire strikes and crashes with incomplete data (Dingus et al., 2006) [40]. Use the SHRP2 data instead [27].

[t] This 81.7% 100-Car study [40] total non-PR crash percentage is incorrectly high because the 82 crash total (note s) includes unreported tire strikes and incomplete data. Use the SHRP2 data instead [27].

[u] Blanco et al. (2016) [27].

[v] See note q.

E.2.1. Errors in Quoting Unreported CV Crash Data

E.2.1.1. Systematic Errors in Abstract

The abstract to Greenblatt et al. (1981) [37] states, "A total of 215 police-reported and 264 non-police reported accidents were described in the form of a detailed set of questionnaires." This abstract contains several systematic errors. First, the Greenblatt et al. (1981) [37] report concerned police-recorded crashes, not police-reported crashes. Second, Greenblatt et al. (1981) [37, p. 3-3] found a total of 315 (54.4%) police-recorded crashes (148 in Phase 1 and 167 in Phase 2), not 215 (a typo in the abstract substituted the digit "2" for "3" as the leading digit). The correct survey figure is 264 (45.6%) unrecorded crashes, which added to the 315 recorded crashes, giving a total of 579 crashes recalled by respondents.

E.2.1.2. Systematic Error in Percentage of Unrecorded Injury Crashes

A later study by Blincoe et al. (2015) [36] found 39.7% unrecorded serious CV crashes (Table E.1, note l). However, three studies, including the Blincoe et al. (2015) [36] study itself, mistakenly cite the earlier Greenblatt et al. (1981) [37] figure of about 24% unrecorded serious crashes:

1. Blincoe et al. (2015, p. 2) [36] state, "Approximately…24 percent of all injury crashes are not reported to the police."
2. Blanco et al. (2016, p. 16) [27] state, "24.3 percent of injury crashes are not reported to police."
3. Schoettle and Sivak (2015, p. 14) [31] state, "24% of injury crashes go unreported each year."

The 24% figure is incorrect because it is at the victim level, not the crash level. It is the percentage of unreported injury victims, not the percentage of unreported crashes in which one or more victims were injured. The correct unreported injury figure at the crash level is 39.7%, as shown in Table E.1 (note l) (see also Blincoe et al., 2015, their Table 1-3) [36].

E.2.1.3. Systematic Error in Percentage of Total Unrecorded Crashes

Blincoe et al. (2015, p. 6) [36] correctly noted a systematic error in the estimate of 81.7% unrecorded crashes (Table E.1, note t) among 82 total crashes (Table E.1B, note s) in the 100-Car NDS [40]. The 81.7% NDS estimate was based on an incorrect count of total crashes. Severity Level 4 incidents such as tire strikes on a curb were erroneously included in the count of 82 total crashes. Since Severity Level 4 incidents by definition are not police reported and are not even considered crashes by many drivers or investigators. The correct number of Level 1-3 total crashes in the 100-Car study is 33 crashes (Table E.1B, notes o,p), of which 63.6% were unrecorded. The SHRP2 study (note u) is a much larger NDS (912 Severity Level 1–3 crashes rather than 33) and should be cited instead.

E.2.1.4. Wrong Number of Households

Blincoe et al. (2015, p. 131) [36] incorrectly claimed that Greenblatt et al. (1981) [37] (Table E.1, note a) "surveyed only 279 households." Greenblatt et al. (1981, p. 3.3) [37] surveyed 11,965 households: 6,541 in Phase 1 and 5,424 in Phase 2. Greenblatt et al. (1981, Table 4-9) [37] show 11,994 total households were phoned.

E.3. Review of Unrecorded CV Crashes

Table E.1 shows a wide range of "Total %" unrecorded crash estimates from 45.6% to 95.0%. On the contrary, 100% of AV crashes in CA are documented in the CA DMV AV database on OL-316 AV crash report forms [32]. Note, however, that extremely few of those AV crashes have an actual police record. Therefore, a direct comparison of AV (all crashes) to CV crashes (only police-recorded crashes) will have a substantial unrecorded crash selection bias. There are two methods to adjust for this bias. First, the CV crash count must have added to it an estimate of police-unrecorded CV crashes. Second, all unrecorded AV crashes must be censored. That is, the AV crashes must be reduced to only police-recorded crashes (see Section 2.4.1, Chapter 2).

The first method is subject to uncertainty because of the wide variation in total unrecorded CV crash percentage estimates from 45.6% to 95.0% (see Table E.1). Hence, any CV crash rate can be made to appear higher or lower than the AV crash rate simply by arbitrarily adding high or low estimates of unrecorded CV crashes. For example, a downward AV IRR bias occurs from adding an estimate of unrecorded CV crashes to the recorded CV crash count from police databases by selecting a study with a high percentage of unrecorded CV crashes. For example, the SHRP2 study in the last row of Table E.1B had a 95% estimate of unrecorded crashes. Conversely, an upward AV IRR bias occurs from adding an estimate of unrecorded CV crashes to the CV recorded crash count in police databases by selecting a study with a low percentage of unrecorded CV crashes. For example, the 1981 survey study [37] in the first row of Table E.1A has a low 45.6% estimate of unrecorded crashes.

Interestingly, the 100-Car NDS (Table E.1B, Lines 1 and 2) [40] and SHRP2 NDS (Table E.1B, Line 3) [27] objectively counted all crashes, as did the AV studies, which altogether avoids the issue of unrecorded crash selection bias. The NDS investigators then diligently searched for police crash reports and records that matched the NDS crashes, even getting assistance after contacting local police departments. As a result, these NDSs found higher percentages of total unrecorded crashes (63.6% and 95.0%, Table E.1B) than did the telephone surveys (Table E.1A). This result suggests that surveys have low-biased estimates of the actual percentage of unreported and unrecorded crashes, consistent with the limitations of survey methods noted by Blincoe et al. (2015, p. 131) [36].

On the other hand, some investigators may believe the survey studies (Table E.1A) are valid and may instead question the high 63.6% to 95% percentages of crashes known to be unrecorded by the police in the 100-Car NDS and SHRP2 NDS (Table E.1B). However, four of the five primary Google contract objectives in the Blanco et al. (2016, p. 1) [27] SHRP2 study were to determine the number and percentages of unrecorded crashes. Therefore, Blanco et al. (2016, their Chapter 3) [27] made a thorough and rigorous effort to ascertain whether every SHRP2 crash had a police report and police record or not. The VTTI investigators carefully examined public police report databases and directly contacted police departments at the six SHRP2 sites to get their assistance in finding matching Police Accident Reports. Blanco et al. (2016, p. 16) [27] note, "a driver may have reported a crash to police or other authorities, but this report was never officially filed, for example: police were called but were unable to respond, or responded but determined that the crash did not meet damage threshold for reporting."

Consistent with the high percentages of unrecorded crashes in the NDSs, this review found that none of the police-reportable AV crashes analyzed by Schoettle and Sivak (2015) [31] had a police record in the state SWITRS crash database [28], even for injury crashes (see Section 2.1.2, Chapter 2).

E.4. Unrecorded AV Crashes

As stated in the first paragraph of the Methods section in Teoh and Kidd (2017) [26], "... the crash rates for Google cars were restricted to those deemed to be police reportable." Note that police reportable is a unique term coined in their study to represent a crash that they judged was likely to be reported to the police. Teoh and Kidd (2017, p. 158) [26] state that "For each crash incident, narratives were used by the authors to code type of collision, injury severity, and whether the crash likely met the California threshold for police crash reporting (fatality, injury, or $750+ in property damage) during the study period."

But it is plausible that almost every AV crash would exceed the CA DMV $750 police-reportable damage criterion because AVs have expensive sensors on their exteriors, where they can be easily damaged in a crash. For example, Teoh and Kidd (2017) [26] classified several crashes as not police-reportable when the property damage likely exceeded the CA DMV reporting requirements (e.g., three Google AV crashes on April 7, 2015; April 27, 2015; and November 2, 2015). Exclusion of such crashes downwardly biased the "police-reportable" autonomous mode crash counts, the AV crash rate, and the AV crash IRR estimates.

Regardless, note that even when Teoh and Kidd (2017) [26] made a "judgment" of a crash as police-reportable, that is not the same as a crash having an actual police record. For example, Teoh and Kidd (2017, Table 2) [26] judged three out of ten crashes in autonomous mode in their Phase 1 (before 2016) were police-reportable (see Table 5.1, note b) and six out of nine as police-reportable in their Phase 2 from January to August 2016. However, this review found no police records in the CA SWITRS [28] police record database for any of these nine crashes that Teoh and Kidd (2017) [26] judged police-reportable, despite injuries, airbag deployments, and tow-aways in some of the crashes.

The following three Sections E.4.1, E.4.2, and E.4.3 are examples of three of the nine crashes that Teoh and Kidd (2017) [26] judged to be police-reportable, none of which had a police record. The first unrecorded crash example is a four-vehicle injury crash with airbag deployment and a tow-away. The second unrecorded crash example is a two-vehicle crash, also with an injury and tow-away. The third unrecorded crash example is a sideswipe with a bus with damage almost certainly exceeding the 2016 CA $750 criterion for filing a crash report with the CA DMV.

E.4.1. Unrecorded AV Crash on July 1, 2015

According to the Google Report for July 2015 [30], the rear-end crash on July 1, 2015, in Mountain View, CA, had a Delta-V of about 17 mph. This crash resulted in whiplash in the "driver, co-driver and rear passenger of the Google AV" that required a hospital visit, while the driver of the other vehicle "reported minor neck and back pain." Waymo (a spin-off company from Google) filed the OL-316 AV crash report [32]. Its crash description is similar, "At the time of the incident, the driver, co-driver, and rear passenger of the Google AV reported some whiplash. Other team members drove them to a local hospital, where they were evaluated by medical staff and cleared to return to work. The driver of the other vehicle reported minor neck and back pain."

The airbag may have deployed in the striking rearward vehicle given the 17 mph Delta-V, which would likely require a tow-away. Yet, Google did not mention an airbag deployment or a tow-away in their Google Report [30] for July 2015 or their DMV OL-316 AV crash report [32] for this July 1, 2015 crash.

However, the airbag deployment was noted by Blanco et al. (2016, Table 5, p. 13) [27]. These investigators report that a Google AV operating in autonomous mode had a rear-end injury crash on July 1, 2015, with a Delta-V of 17 mph, a high peak acceleration of 3.5 |g|, moderate damage to the other vehicle's front end, airbag deployment, and a tow-away. Blanco et al. (2016, Table 5, p. 13) [27] explicitly state that there was "visible deployment of the driver-side front airbag in the following vehicle." In addition, the VTTI investigators confirmed the whiplash injuries of the three Google AV occupants and the neck and back pain of the driver of the other vehicle, consistent with the OL-316 AV crash report [32].

Note also that this was a multi-vehicle crash involving four vehicles, as mentioned in the Google Report [30] for July 2015 and the CA DMV OL-316 AV crash report [32] for July 1, 2016, filed by Waymo, but not in the crash description by Blanco et al. (2016) [27].

CBS News [41] reported that "Mountain View police responded, but did not file an accident report," which was not mentioned by Blanco et al. (2016) [27], the Google Reports [30] for July 2015, or the DMV OL-316 AV crash report [32] filed by Waymo for this July 1, 2015, injury crash. In the additional analysis in this critical review, there was no police record for this multi-vehicle injury crash in the SWITRS database [28], accessed through TIMS [29].

E.4.2. Unrecorded AV Crash on August 20, 2015

This crash resulted in a back injury requiring a hospital visit. In Mountain View, CA, a Google rear-end injury crash had a Delta-V of 5 mph, peak acceleration of 2.3 |g| (Blanco et al., 2016, Table 5, p. 13) [27] and a tow-away of the other vehicle. The Google DMV OL-316 AV crash report [32] and Blanco et al. (2016, their Table 5, p. 13) [27] state, "The Google AV test driver reported minor back pain and was taken to a local hospital by Google employees, where he was evaluated and released by medical staff." Also, "The other vehicle sustained moderate damage to its front end and was towed from the scene" (Blanco et al., 2016, Table 12, p. 50) [27]; Google Reports, August 2015 [30]; Google DMV OL-316 AV crash report [32]). (For further details, see the crash description in the Google Report [30] for August 20, 2015, quoted in Appendix C.) A police record for this injury crash was not in the SWITRS database [28], accessed through TIMS [29].

E.4.3. Unrecorded AV Crash on February 4, 2016

A Google Lexus AV had a sideswipe collision with a public transit bus in Mountain View, CA. "The Google AV sustained body damage to the left front fender, the left front wheel, and one of its driver's-side sensors" (Google DMV OL-316 AV crash report [32]). This amount of damage almost certainly exceeded the 2016 CA $750 requirement to file a "Report of Traffic Accident Occurring in California" (SR 1) with the DMV. Nonetheless, a police record for this likely Major PDO crash was not in the SWITRS database [28], accessed through TIMS [29].

E.5. Discussion of Appendix E

Why do so few, if any, AV crashes in the CA field test have a formal police record in the CA SWITRS [28] police record database? This paucity holds even for some Severe crashes involving an injury, airbag deployment, and tow-away. Some possible reasons are suggested below.

A. AV company reasons

1. The AV company prefers not to report AV crashes to the police.
2. The AV company does not consider its prototype AV test vehicles subject to the same police reporting criteria as CVs, even for Severe crashes (e.g., when emergency vehicles or tow trucks were called to the scene, airbags were deployed, or one or more victims were injured).
3. The AV company believes it is redundant and unnecessary to file a police report because they intended to mention the crash in a separate press release or report the crash on their website (e.g., Google Reports [30]).
4. For crashes after the date of issuance of the Manufacturer's Testing Permit to an AV company (e.g., Google's issuance date was September 24, 2014), all AV crashes on CA public roads had to be reported by the AV company to the CA DMV AV database within ten days, or the company would lose its AV testing license. These AV crashes are placed in a separate DMV database intended specifically for AV crash reports [32], a different database from the SWITRS [28] police record database. Thus, an AV company might believe it is unnecessary to make a separate crash report to the police because it was already filing a mandatory OL-316 AV crash report in the CA DMV AV database [32].

B. Police reasons

Another set of possible reasons occur in crash cases where someone other than an AV company or its test driver or occupants in the AV contacts the police. Or, in some cases, the police were passing by and stopped at the crash scene, etc. The police were informed that the crash occurred in such cases but chose not to write a Police Accident Report or file a formal police record in the CA SWITRS [28] database. The following three additional reasons provide possible explanations:

1. The police already knew, or might be informed by AV company representatives if the police did not know, that the AV company must separately report all AV crashes to the DMV. Therefore the police may have judged it unnecessary and redundant to file a separate crash record in the SWITRS [28] police database.
2. In the case of Google/Waymo, whose ODD is mainly in Mountain View, the Mountain View Police Department has less of a tendency to file a formal police record in the SWITRS [28] database than other police jurisdictions. Indeed, as mentioned in Section 5.1.2.1 and Appendix E, the police criteria to file a formal record varies considerably in different municipalities, particularly for PDO crashes without injuries. For example, the Mountain View Police Department does log many bench reports after crashes (see Mountain View Police Department Annual Reports, 2009–2017 [42]). However, an investigation by this review found that these crashes rarely have a formal police record filed for them in the CA SWITRS [28] database.
3. For some crashes, it is stated in the OL-316 AV crash report [32], "Police were notified of the incident," such as a Google AV Major PDO crash in Los Altos, CA, on July 15, 2016. However, the Los Altos Police Department did not file a record in the state SWITRS (2016) [28] police database.

See Appendix D for further discussion about OL-316 AV crash reports [32] and other crash descriptions by AV companies.

The bottom line is issues concerning unreported and unrecorded crashes are not settled in the crash safety literature. This uncertainty makes the comparisons of AV crash rates to CV crash rates based on crashes in police record databases highly problematic. Studies that record all CV crashes, such as an NDS, offer a better CV comparison benchmark, but even there, there are site selection bias issues (see Section 3.3 in the main body).

An optimum benchmark for the CV crash rate avoids both unrecorded crash selection bias and site selection bias. The optimum benchmark is to estimate the comparison CV crash rate from the crashes and mileage of autonomous vehicles operating in manual mode. This comparison has a similar ODD, and all crashes would be documented. Unfortunately, although the manual mode crashes are listed in the CA OL-316 AV crash reports [32], the mileage while operating in manual mode is not provided in the CA disengagement reports [43]. As a result, it is impossible with current publicly available data to estimate a CV crash rate using the optimum benchmark of manual mode crash and mileage with the identical vehicles that operate in autonomous mode.

A valid estimate of unreported and unrecorded crashes is also critical for estimating total crash costs for the United States. Three NHTSA U.S. crash cost studies were based on the Greenblatt et al. (1981, Tables 4-4 and 5-2) [37] estimated percentages of unrecorded crashes in Table E.1A (note c). These cost studies were Blincoe and Faigin (1992, their Table III-38) [44], Blincoe (1996, p. 9) [45], and Blincoe et al. (2002, p. 9) [46]. The estimated percentages of unrecorded crashes they used are listed in Table E.1A.

One birds-eye observation from Table E.1 is that in every study of the unreported and unrecorded crashes so far, the estimated percentages of unrecorded PDO crashes are higher than the estimated percentages of unrecorded injury crashes. That observation is plausible because it is intuitively apparent that people injured in a crash are more likely to contact the police. Likewise, the police are more likely to file a record in state police databases for an injury crash than a PDO crash.

Another birds-eye observation from Table E.1 is that the estimated total percentages of **unrecorded** crashes (in **bold**) are all higher than the estimate of the total percentage of *29.3% unreported* crashes (in *italic*). This observation is also intuitively apparent because unreported crashes cannot be recorded, and many reported crashes are never recorded (see Appendix E).

Appendix F
Are AVs or Human Drivers at Fault in a Crash?

CONTENTS

Note: Refer to Appendix B on page 149 for an explanation of the abbreviations and the definitions of unique terms and phrases.

F.1. Review of Fault Claims

For crashes in which an AV was involved, three of the six reviewed studies stated that AVs were not at fault; the human driver was. Two studies and one newspaper report stated that the AV was at fault in one crash, the human driver was in all others. One study made no statement regarding fault for either AVs or human drivers. Details are below.

A. Three studies claimed that the AV was not at fault and human drivers were at fault in all crashes involving an AV in their study period.

1. Schoettle and Sivak (2015) [31]. Based on the crash narratives supplied by the companies [30, 32], this study [31, p. 6] stated, "Self-driving vehicles have not been found to be at fault in any of the 11 crashes occurring in autonomous mode." The study claimed that this is a "fact" and said it "is consistent with the anticipated uncertainty about what to expect from self-driving vehicles on the part of drivers of CVs." However, the study presented no evidence to substantiate its conclusion that AVs were not at fault in any crashes in which they were involved. Moreover, no statements by the other drivers were included in the public crash descriptions by the AV companies [30, 32]. Therefore, it is unclear on what basis this study claimed that the CV drivers had "uncertainty about what to expect" of AV behavior.

2. Blanco et al. (2016) [27]. The internal Google AV test data was made available to these investigators, including videos of the crashes. Based on that data, Blanco et al. (2016, p. 41) [27] state, "When the Self-Driving Car events were analyzed using methods developed for SHRP2, none of the vehicles operating in autonomous mode were deemed at fault." The study states that the same criteria were used in SHRP2 for assigning fault to the Google AV crash cases. In the SHRP2 study, "fault" was assigned for the CVs according to "which driver or non-motorist (if any) committed an error that led to the event." However, the criteria used in the SHRP2

study to make a judgment about "which driver committed an error that led to the event" were not presented in SHRP2 reports, as discussed in [27], other than to state that the assessment was based on a review of the crash video.

3. Favarò et al. (2017) [33]. This study stated in its Section 3.3 (p. 15), "A careful analysis of the accidents' descriptions shows that in 22 out of the 26 reported accidents the AV was not-at-fault." In the four cases in which the AV was said to be at fault, two happened during the manual mode. In the other two AV at-fault cases, the AV was not in autonomous mode, meaning the ADS was disengaged during the crash sequence. So this study concluded that the AV was never at fault while in autonomous mode.

B. Two studies and one newspaper report claimed that the AV was at fault in one crash and human drivers in the rest.

1. Dixit et al. (2016) [13]. This study stated (p. 6) that from September 2014 to February 2016, "Out of the twelve reported accidents, only one of them involved the autonomous vehicle being at fault. This was attributed to the autonomous vehicle incorrectly expecting the other vehicle to give way." The one crash where the AV was deemed at fault in this study was a sideswipe crash with a bus on February 14, 2016 (Bhattacharjee, 2016) [47]. Dixit and colleagues claimed that the AV had an incorrect "expectation." The study further claimed that "In most of the cases where the other vehicle was determined to be at fault, the underlying cause could be attributed to the driver of the other vehicle expecting the autonomous vehicle to behave differently from what they would have normally expected." However, no criteria or evidence are given about how this study decided that the human driver was at fault in the other 11 AV crashes. Also, it is not clear from what evidence this study drew its conclusions about the "expectations" of the CV drivers or the "expectations" of the AV. It is also not plausible that a neural network ADS can have "expectations."

2. Teoh and Kidd (2017) [26]. This study stated (p. 60) that "…the Google car was partially at-fault in only one incident." This study provides no evidence or criteria for that statement, nor on what basis the authors judged the human driver to always be at fault in the other AV crashes that were investigated.

3. Kokalitcheva (2017) [48]. The reporter for Axios investigated 24 CA DMV OL-316 AV crash reports [32] indicating autonomous mode from 2014 to 2017. In only one crash did the reporter state that an AV was "at fault." The reporter further stated that "Axios had to make assumptions in a few cases, including the car's general speed and which driver was at fault, due to information gaps in the accident reports." The report concluded that "humans cause most self-driving car crashes."

C. One study did not mention fault.

1. Banerjee et al. (2018) [22]. This study mainly analyzed AV failures leading to disengagements, with only a small section on crashes. These researchers made no statements about whether the AV or human driver was or was not assigned fault in crashes involving an AV.

In summary, three studies assigned fault to the human driver in all AV crash cases: Schoettle and Sivak (2015) [31], Blanco et al. (2016) [27], and Favarò et al. (2017) [33]. Two studies (Dixit et al., 2016 [13] and Teoh and Kidd, 2017 [26] and one newspaper report (Kokalitcheva, 2017) [48] assigned fault to the human driver in all AV crash cases but one. One study did not mention fault (Banerjee et al., 2018 [22]).

F.2. Fault Issues

Five scientific studies (Dixit et al., 2016 [13], Teoh and Kidd, 2017) [26], Blanco et al., 2016 [27], Schoettle and Sivak, 2015 [31], Favarò et al., 2017 [33]) and one newspaper report (Kokalitcheva, 2017 [48]) contained the following issues regarding fault assignment.

1. The studies did not describe their criteria for assigning fault and presented no evidence for those fault assignments.
2. The studies took the crash descriptions (Google Reports [30], CA DMV [32]) at face value (see Appendix D).
3. The authors of the studies may have been influenced by Google's public statements to news media and websites that it was the human driver's fault in almost all AV crashes. Google has claimed in their public reports and statements (see LaFrance, 2019 [51]) that Google AVs have rarely been at fault in autonomous mode in any Google crash during the period investigated in these studies. The one exception that Google acknowledged was the bus crash on February 14, 2016 (Google Reports [30]). But, again, Google did not state its criteria for assigning fault, nor did it present evidence to support its assignment of fault to the human driver in every crash case except one. Google even claimed, without evidence, that the human drivers crashed into the Google AVs because of driver distraction, which Google said enhanced its case for autonomous driving.
4. The word "fault" is not present in any of the crash descriptions in the AV monthly crash reports in Google Reports [30] or its crash reports to the CA DMV [32] in the period studied by these investigators.
5. There is no "at fault" category in the CA DMV OL-316 reports [32], unlike Police Accident Reports.
6. AV company reports [30, 32] give only the company's version of the crashes. Comments from drivers or passengers of other vehicles, or nearby witnesses, were not present in any of the Google Reports [30] or any OL-316 AV report from any AV company [32].

F.3. Discussion of Fault Issues

The five studies [13, 26, 27, 31, 33] and one newspaper report [48] that assigned fault failed to specify the criteria or evidence used for making those fault assignments. None of the studies mentioned searching for independent public evidence, such as police reports or interviews with the other vehicle's driver, or witnesses in newspaper reports. Favarò et al. (2017) [33] explicitly state that they relied solely on the AV companies' crash descriptions. Thus, the claim of no AV at fault in any crash, except one, was not supported with any objective, independent evidence.

No study acknowledged that public comments by AV developers would likely minimize crash severity and their potential fault for liability purposes. In addition, no study acknowledged that a determination of fault in a crash could be a contentious issue for insurance companies. Or in the legal system, particularly when liability lawyers become involved. Fault becomes an even more contentious issue if a legal case reaches a courtroom, although most automotive liability cases are settled out of court. See Section F.5 for a further discussion of legal issues concerning fault assignment in a multi-vehicle crash.

This review read the same CA DMV OL-316 AV crash reports [32] and Google Reports [30] used by the authors of the reviewed scientific studies. It is unclear from the information in those crash reports how the reviewed studies could have made an objective determination of whether the human driver or the AV was at fault in a crash involving an AV. In other words, based on the facts as presented in Section F.1, it is not clear on what evidence the studies assigned fault to either the AV or the human driver.

F.4. Recommendations about Fault Issues

The following fault recommendations are made to authors of scientific articles about AV safety:

1. Do not assign fault based solely on crash descriptions by AV companies in public reports [30, 32].
2. Do not assign fault based solely on public statements by AV companies.
3. If recommendations one and two are ignored (and the fault is assigned to an AV or human in an AV crash), then acknowledge in a limitations section that a substantial limitation of the study is that the fault assignment was a subjective opinion without evidence.

F.5. Legal Questions regarding AV Fault

1. Consider the common scenario where a human driver in a CV rear-ends an AV at an intersection. AV advocates would claim that the fault lies with the human driver, not the AV. Perhaps the AV advocates simply apply the common legal criterion for human-driven CVs. Namely, a driver who rear-ends another vehicle is at fault unless the lead driver is breaking the law. At busy intersections, Google Reports [30] and Google CA DMV reports [32] state in multiple crash descriptions that the Google AVs sometimes "crept" forward. Such multiple start-and-stop vehicle behavior presumably occurs as the ADS inspects the oncoming traffic, trying to determine when it is safe to turn at the intersection. Similar AV behavior might occur at four-way stops.
2. Consider another common scenario in everyday driving with CVs that might elicit the "creeping" AV behavior. The forward vehicle is stopped, looking for an opportunity to merge with oncoming traffic. The forward vehicle finally starts moving. In ordinary everyday driving experience, the driver who immediately follows takes that forward movement of the lead vehicle as the cue to begin advancing. The following driver would then typically scan to the side for oncoming traffic to estimate a gap large enough to merge with oncoming traffic safely. (It is necessary for safe driving to briefly take the eyes off the forward roadway to check for traffic from the side before pulling out to merge with traffic.) Now assume that

the forward vehicle suddenly stops because the forward driver reconsiders and decides it is not safe to merge. At the same time as the forward vehicle suddenly stops, the following driver's eyes are still off the forward roadway. The following driver might not see such a sudden stop of the forward vehicle with enough time to stop after scanning to the side and returning their gaze to the forward roadway. With human drivers in the two vehicles, insurance companies would have to battle out which insurance company has liability in this rear-end crash scenario. If a severe injury or a death is involved, the case might go into the legal system. Either way, it will be a long and contentious battle to decide who was or was not at fault.

3. It is not clear how in such cases (or many others), a scientific study can easily assign fault when it can be a contentious and protracted legal battle. Because of this contentiousness, 12 states have adopted "no-fault" auto insurance laws (Insurance Information Institute, 2018) [52]. However, insurance companies and courts can still end up in difficult decisions about the relative degree of fault of the parties involved in a crash, particularly if deaths or severe injuries are involved.
4. When an AV is the lead vehicle, and a rear-end crash happens in these scenarios, are the liability questions the same or not? AVs can detect vehicles behind them better than humans because AVs have rear-facing sensors, which humans do not. Does that increase the AV company's liability or not for CV crashes into the rear of their AV? Suppose the CV driver who rear-ended the AV was interviewed or gave a deposition in a legal case. Assume that the AV was the forward vehicle and attempted to merge in the on-ramp scenario. Assume also that it exhibited a cautious stop-and-go behavior (i.e., it "crept" forward). Would the following human driver (or his insurance company or lawyer) maintain that the AV was at fault for suddenly stopping after it had started moving forward on the on-ramp?
5. As noted above, comments from drivers of other vehicles are not included in any Google Reports [30] or OL-316 AV report from any AV company [32]. Without this information, the fault assignment could be wrong. One example is a crash in San Francisco between a GM Cruise AV and a motorcycle on December 7, 2017, which injured the motorcyclist. In its report to the CA DMV [32], the GM Cruise crash description did not include any statement from the motorcyclist and claimed that a San Francisco police department "Traffic Collision Report" showed it was not at fault. Using the methods of the scientific studies reviewed here, they would likely claim (as did the AV developer) that the AV was not at fault; the motorcyclist was at fault. However, a newspaper account by Baron (2018) [49] states that the motorcyclist sued GM Cruise in what became the first AV liability case. The motorcyclist claimed that the AV was at fault. The case was eventually settled out of court (Felton, 2018) [50]. (See Section C.1.2 in Appendix C for a further discussion of this December 7, 2017, GM Cruise/motorcycle crash case.)

Appendix G
Crash Severity Definitions in Naturalistic Driving Studies

Note: Refer to Appendix B on page 149 for an explanation of the abbreviations and the definitions of unique terms and phrases.

TABLE G.1 Crash Severity Level definitions in the 100-Car NDS (Dingus et al., 2015) [40] and SHRP2 NDS (Blanco et al., 2016) [27].

Severity	100-Car[a]	SHRP2[b]
Level 1	"Police-reported air bag deployment and/or injury."	"Crashes with airbag deployment, injury, rollover, a high Delta-V, or that require towing. Injury, if present, should be sufficient to require a doctor's visit, including those self-reported and those from apparent video. A high Delta-V is defined as a change in speed of the subject vehicle in any direction during impact greater than 20 mph (excluding curb strikes) or acceleration on any axis greater than ±2 g (excluding curb strikes)."
Level 2	"Police-reported property damage only."	"Crashes that do not meet the requirements for a Level 1 crash. Includes sufficient property damage that one would anticipate is reported to authorities (minimum of $1,500 worth of damage, as estimated from video). Also includes crashes that reach an acceleration on any axis greater than ±1.3 g (excluding curb strikes). Most large animal strikes and sign strikes arc considered Level 2."
Level 3	"Non-police-reported property damage only."	"Crashes involving physical conflict with another object (but with minimal damage) that do not meet the requirements for a Level 1 or Level 2 crash. Includes most road departures (unless criteria for a more severe crash are met), small animal strikes, all curb and tire strikes potentially in conflict with oncoming traffic, and other curb strikes with an increased risk element (e.g., would have resulted in a worse crash had the curb not been there, usually related to some kind of driver behavior or state, for example, hitting a guardrail at low speeds)."
Level 4	"Non-police-reported low-g physical contact or tire strike (greater than 10 mph)."	"Tire strike only with little or no risk element (e.g., clipping a curb during a tight turn). Distraction may or may not also be present. The distinction between Level 3 and Level 4 crashes is that Level 3 crashes would have resulted in a worse crash had the curb not been there while Level 4 crashes would not have due to the limited risk involved with the curb strike."

Notes:

[a] Dingus et al. (2015, p. xxxvi) [40].

[b] Blanco et al. (2016, pp. 3–4) [27].

Appendix H
Standardization Method

Contents

Note: Refer to Appendix B on page 149 for an explanation of the abbreviations and the definitions of unique terms and phrases.

H.1. Rationale for AV Crash Site Standardization

The Blanco et al. (2016) [27] study was reviewed in Chapter 3. It compared Google AV crash rates to SHRP2 CV crash rates, stratified by crash Severity Level. Figure 3.1 shows the study's original Google AV crash rates in Mountain View and vicinity compared to the SHRP2 CV crash rates at the SHRP2 sites. Figure 3.2 shows the Google AV crude IRR estimates from the crash rates in Figure 3.1. Figures 3.1 and 3.2 suggest that Google AVs reduce CV crash rates at every crash Severity Level.

However, these apparent reductions could be due to site selection bias. This bias arises from different CV crash rates at the two sites. The size and direction of the bias might also vary by crash Severity Level. That is, the appearance in Figure 3.1 that Google AVs have a lower crash rate than CVs might be more simply explained if there is just a generally lower crash rate in Mountain View and vicinity than at the SHRP2 sites. The crash rate reductions are not caused by Google AVs but simply by the lower crash rates at the test site selected by Google. Such an interaction between the AV effect and the site effect is here called *site selection bias*.

H.1.1. Google AV Severe Crash Rate Reduction Is a True Effect

Figure 3.3A shows that the Severe crash rate estimate of Google AVs in the Mountain View area was 1.6 (green bar). This Google AV rate was 36% lower than the SHRP2 CV Severe crash rate estimate of 2.5 (blue bar) at the SHRP2 sites.

Figure 3.3B shows that the CV police-recorded injury crash rate estimate in Mountain View was 0.68 (right bar). That rate was 14.7% higher than the SHRP2 CV Severe crash rate estimate at the SHRP2 sites of 0.58 (left bar) during the same period. Therefore, the Google AV 36% Severe crash rate reduction effect in Figure 3.3A cannot result from site selection bias because Mountain View does not have a relatively low CV injury crash rate. The police-recorded Mountain View injury crash rate was higher than the PR injury crash rate for the SHRP2 sites over the same period. The 36% Severe crash rate reduction by Google AVs versus SHRP2 CVs is the opposite of the 14.7% higher injury crash rate in Mountain View versus the SHRP2 sites. It follows that the Google AV Severe crash rate reduction effect of 36% is even more substantial than suggested by the crude rate comparison in Figure 3.3A. The reason is that the 36% reduction occurred despite the 14.7% higher injury crash rate expected in Mountain View versus the SHRP2 sites.

Note that the absolute values of the crash rates in the SHRP2 and Google AVs in Figure 3.3A are far higher than the absolute values of the PR crash rate at the SHRP2 and police-recorded crash Mountain View Sites in Figure 3.3B. That is because the SHRP2 study and the Google AV study recorded all crashes. The PR crashes are as low as 5% of all crashes in the SHRP2 study estimate (see Appendix E). That is why the two plots in Figure 3.3A and B are done with separate scales, and only the trends are considered. An absolute comparison between crash rates for all crashes versus police-recorded or PR crashes is subject to unrecorded crash selection bias and is avoided using the trend analysis method.

H.1.2. Google AV PDO Crash Rate Reductions Are Not True Effects

Figure 3.4A shows that the Major PDO crash rate estimate of Google AVs in the Mountain View area was 1.6 (green bar). This Google AV rate was 51.5% lower than the SHRP2 CV Major PDO estimate at the SHRP2 sites of 3.3 (blue bar), suggesting that Google AVs reduce Major PDO crash rates compared to human drivers.

However, Figure 3.4B shows a 52.9% lower Major PDO crash rate for CVs in Mountain View than at the SHRP2 sites. This result suggests that the Google AV's apparent reduction in Major PDO crash rates from 3.3 to 1.6 seen in Figure 3.2 (middle bars) is not a true effect. The 51.5% reduction by Google AVs was in the same direction and size as the 52.9% reduction in police-recorded Major PDO crash rates during the same period. Thus, the apparent Google AV Major PDO crash rate reduction effect is more simply explained by the fact that Google selected Mountain View as its primary AV test site, and that site has a lower Major PDO crash rate than the SHRP2 sites.

Regarding Minor PDO crash rates, the SHRP2 Minor PDO crash rate was 14.4 in the SHRP2 study (light blue bar to the right of Figure 3.1). The Google AVs had a substantially lower Minor PDO crash rate of 5.6 (rightmost bar in Figure 3.1). The SHRP2 study did not count PR Minor PDO crashes, so an estimate cannot be made for police-recorded Minor PDO crashes at the SHRP2 sites. However, it is plausible that the Minor PDO crash rate is far lower in Mountain View than at the SHRP2 sites, as it was for Major PDO crash rates. In that case, the Google AV reduction in the Minor PDO crash rate compared to the SHRP2 Minor PDO crash rate is biased low by site selection bias, as was the Major PDO crash rate.

H.1.3. Benefits of Standardization

However, Figures 3.3B and 3.4B are just qualitative demonstrations of the effects of site selection bias. The standardization method adjusts the AV crash rate quantitatively for these differences in the CV crash rate at the two sites. Standardization thus allows for a valid quantitative comparison of the Google AV crash rates to the SHRP2 CV crash rate at every crash Severity Level. In other words, the standardization adjusts the AV crashes and AV crash rate for the site selection bias so that a more valid AV IRR estimate can be made, with site selection bias removed. The final standardized count of AV crashes is the number of AV crashes predicted to occur if the Google AV tests had been conducted at the SHRP2 sites instead of at the Mountain View and vicinity site.

Standardization adjusts the AV crash counts to eliminate site selection bias by comparing the CV crash rates at the two sites. Therefore, the differences in the CV crash rates at the two sites, as shown in Figures 3.3 and 3.4, are the data used to standardize the Google AV crash counts in Mountain View to those at the SHRP2 sites.

In short, site selection bias in the Google AV crash rate estimates by Blanco et al. (2016) [27] was identified in Section 3.2.2. The bias was removed by standardizing the Google AV crash rates at the Mountain View site to the SHRP2 CV sites. The standardization factors are the ratios of the CV crash rates between Mountain View and the SHRP2 CV sites at each crash severity, as shown in Figures 3.3B and 3.4B. The AV crash counts that give rise to the AV crash rates in Figures 3.1, 3.3A, and 3.4A were standardized using these standardization factors. The details of the calculation methods are illustrated in the following section.

H.2. Standardization Calculation Method

The standardization calculation method is illustrated in Table H.1.

H.2.1. AV Severity Level 1 (Severe) Crash Standardization Factor

The data for Level 1 (Severe) crash rates in Table H.1 indicate that the ratio of the CV injury crash rate SHRP2 sites (Var. #1) to the CV injury crash rate at the Mountain View site (Var. #2) is 0.58/0.68 = 0.853 (Var. #3). This ratio is the standardization factor for Level 1 (Severe) Google AV crashes. Next, multiply the 0.853 standardization factor (Var. #3) by the Google AV Level 1

TABLE H.1 Method to standardize AV crashes and rates for the Blanco et al. (2016) [27] data.

			Crash Severity Level					
Variable	Var. #	Calc.	Level 1 (Severe)	Source	Level 2 (Major PDO)	Source	Level 3 (Minor PDO)	Source
SHRP2 6 sites CV PR crash rate	1		0.58	Fig. 3.3B, left	1.21	Fig. 3.4B, left	N/A	N/A
Mountain View CV police-recorded crash rate	2		0.68	Fig. 3.3B, right	0.57	Fig. 3.4B, right	N/A	N/A
Standardization factor	3	(1 ÷ 2)	0.853		2.123		2.123	Major PDO
Google crude AV crashes	4		2	Tbl. 3.1A, note b	2	Tbl. 3.1B, note b	7	Tbl. 3.1C, note b
Standard Google AV crashes	5	(3 × 4)	1.71	Tbl. 3.2A, note b	4.25	Tbl. 3.2B, note b	14.86	Tbl. 3.2C, note b
Google AV MVMT	6		1.20988	Tbl. 3.2A, note c	1.20988	Tbl. 3.2B, note c	1.20988	Tbl. 3.2C, note c
Standard Google AV crash rate	7	(5 ÷ 6)	1.41	Fig. 3.5, left grn	3.51	Fig. 3.5, middle grn	12.28	Fig. 3.5, right grn
95% LL	8		0.277	Epi. [53]	1.05	Epi. [53]	7.20	Epi. [53]
95% UL	9		5.46	Epi. [53]	7.97	Epi. [53]	20.0	Epi. [53]

(Severe) crashes (Var. #4). The product is the estimated number of Level 1 (Severe) crashes that would be expected if the Google AV test at the Mountain View site had been conducted at the SHRP2 NDS sites (Var. #5), or 1.71 crashes. Dividing this standardized Google AV crash count (Var. #5) by the observed Google AV mileage of 1.20988 MVMT (Var. #6) estimates a standardized Google AV Severe crash rate of 1.41 Severe crash per MVMT (Var. #7). The mid-P exact 95% confidence limits (0.277, 5.46) (Vars. #8, #9) of the standardized Google AV Severe crash rate (Var. #7) were then estimated using the Episheet [53] Quickcalc tab.

H.2.2. AV Severity Level 2 (Major PDO) Crash Standardization Factor

The data for Major PDO crashes in Table H.1 indicates that the ratio of the CV PR crash rate at the SHRP2 sites (Var. #1) to the CV police-recorded crash rate at the Mountain View site (Var. #2) is 1.21/0.57 = 2.123 (Var. #3). This ratio is the standardization factor for Google AV Major PDO crashes. Next, multiply the Google AV Major PDO crash count of 2 (Var. #4) by the Major PDO standardization factor of 2.123 (Var. #3). The product estimates the standardized count of Google AV Major PDO crashes of 4.25 crashes (Var. #5). Dividing this Google AV Major PDO standardized crash count by the Google AV mileage of 1.20988 MVMT (Var. #6) estimates a standardized Google AV Major PDO crash rate of 3.51 crashes per MVMT (Var. #7). The mid-P exact 95% confidence limits (1.05, 7.97) (Vars. #8, #9) of this standardized Google AV Major PDO crash rate were then estimated with the Episheet [53] Quickcalc tab.

H.2.3. AV Severity Level 3 (Minor PDO) Crash Standardization Factor

In the SHRP2 study, PR crashes were excluded in the Minor PDO crash stratum by definition, but PR crashes were included in the Major PDO stratum. Appendix G shows this exclusion was done by definition in the 100-Car study [40] that preceded the SHRP2 study. But even though it is not described in the SHRP2 crash severity definitions (Appendix G), it was nonetheless the practice in the SHRP2 study to exclude PR crashes from the Minor PDO crash stratum.

This exclusion has a consequence for the estimate of the site standardization factor for Minor PDO crashes. As noted in Section H.1.2, there were no PR Minor PDO crashes in the SHRP2 study by definition. Hence, it was impossible to calculate the standardization factor by comparing the police-recorded Minor PDO CV crash rate at the Mountain View site to the PR Minor PDO CV crash rate at the SHRP2 sites because the latter was not defined. As a workaround for this missing SHRP2 PR Minor PDO crash data, it was assumed that the Major PDO standardization factor of 2.123 (Var. #3) was also a plausible estimate for the standardization factor to adjust Minor PDO crashes (in the absence of any other information). To standardize the Google AV Minor PDO crash site in Mountain View to that in the SHRP2 NDS sites, the Google AV Minor PDO crash count of seven (Var. #4) was multiplied by the standardization factor 2.123 (Var. #3). Dividing by the Google AV mileage (Var. #6) then estimates the standardized AV Minor PDO crash rate of 12.28 (Var. #7). The mid-P exact 95% confidence limits (7.20, 20.0) (Vars. #8, #9) of the AV Minor PDO crash rate were then estimated with the Episheet [53] Quickcalc tab.

Appendix N discusses a PR crash selection bias that remains in the standardized Google AV crash rate in Table 3.2 and Figure 3.5, and the standardized Google AV crash IRR in Table 3.2 and Figure 3.6.

Appendix I
Replication of Teoh and Kidd (2017) Phase 1 CV Crash Results (Table 12, Note f)

Note: Refer to Appendix B on page 149 for an explanation of the abbreviations and the definitions of unique terms and phrases.

Teoh and Kidd (2017, Table 3) [26] tabulated 9,198 passenger vehicles (see Chapter 5, Section 5.1.1.5, Table 5.1, note f) involved in police-recorded crashes in Mountain View city streets and highways within Mountain View city limits from 2009 to 2015.

To replicate this count, a query was sent to the CA SWITRS police-record database [28] for all the raw data for the police records in Mountain View from 2009 to 2015. This request results in a Collision file, a Party file, and a Victim file. The Party file tabulates the types of vehicles involved in particular crashes. Table I.1 shows vehicle types involved in crashes using the Party File information in the SWITRS database [28] for Teoh and Kidd's (2017) [26] Phase 1 period.

Teoh and Kidd (2017) [26] did not define what they meant by the term "passenger vehicles." But Table I.1 shows that the sum of the crash-involvement vehicle count for the four vehicle types closely replicated the 9,198 total in Teoh and Kidd (2017, Table 3) [26]. The difference was only six vehicles (0.07%). The discrepancy of six vehicles may have been the result of updates by the CHP of the SWITRS database [28] after the date that Teoh and Kidd (2017) [26] requested the data from that database.

Table I.1 indicates that Teoh and Kidd (2017) [26] likely used the term "passenger vehicles" as a generic term. The term included the vehicle types labeled A, B, D, and E in Table I.1, even though subtypes D and E are pickup or panel trucks and not passenger vehicles.

How did the police count these vehicles when they filed the police records in the state SWITRS database [28]? In other words, what vehicle subtypes did the police include in the vehicle categories "passenger car" and "pickup or panel track" in Table I.1? A plausible answer is given by the CA DMV (2021) [54], wherein a passenger vehicle is defined as an automobile that "does not transport persons for hire. This includes station wagons, sedans, vans, and sport utility vehicles."

In short, Teoh and Kidd (2017) [26, their Table 3] mistakenly labeled their 9,198 total as a crash count, but it was, in fact, a count of vehicles involved in crashes, as demonstrated by Table I.1. In other words, Teoh and Kidd's (2017) [26] crash-involved count of 9,198 (Chapter 5, Table 5.1, note f) for Mountain View is at a vehicle level, not a crash level. Thus, for example, a CV crash involving two passenger vehicles was counted as two crash-involved vehicles by Teoh and Kidd (2017) [26], not one crash. Similarly, a CV crash with three passenger vehicles involved was counted as three crash-involved vehicles, not one crash, and so forth. In other words, for their CV count, Teoh and Kidd (2017) [26] tabulated the total number of vehicles involved in the police record in the SWITRS [28] database, not the number of CV crashes. Thus, using this vehicle-level definition, Teoh and Kidd

TABLE I.1 Replication of Teoh and Kidd (2017) CV crash-involvement count: Counts of vehicle types involved in Teoh and Kidd's (2017) Phase 1 period according to the State of CA police records.

Vehicle type	Count
A - Passenger car/Station wagon	8,379
B - Passenger car with trailer	27
D - Pickup or panel truck	767
E - Pickup or panel truck with trailer	31
Total	9,204

Counts of crash-involved human-driven vehicle types from 2009 to 2015 in Mountain View, CA.

(2017, Table 3) [26] tabulated 9,198 human-driven passenger vehicles involved in police-recorded crashes in Mountain View city streets and highways within Mountain View city limits from 2009 to 2015.

The number of vehicles involved in a crash is contained in the "Party" record for a crash in the SWITRS [28] database. By finding the unique police record ID numbers for a crash within the Party File database, the actual number of passenger vehicle police-recorded crashes in city streets and highways within Mountain View city limits from 2009 to 2015 was determined to be only 4,862, not 9,198. After eliminating other errors and biases in Teoh and Kidd's (2017) [26] analysis (see Section 5.1.2), the actual number of CV crashes fell to 1,993 as per Table 5.2 (note f) in Section 5.1.3.

Teoh and Kidd's (2017) [26] method of counting the number of CVs involved in crashes rather than CV crashes is another form of crash selection bias because they used a different criterion to tabulate the AV group than the CV group. For example, Teoh and Kidd (2017) [26] counted only three AV crashes (in crashes that they judged as police-reportable), even though they were multi-vehicle crashes, as follows.

1. On July 1, 2015, an injury crash involved four vehicles: one AV and three CVs (see Section E.4.1 for description). Teoh and Kidd (2017) [26] added only one crash to their AV crash count, not four.
2. In March 2013, a Major PDO crash involved two vehicles: one AV and one CV. Teoh and Kidd (2017) [26] added only one crash to their AV crash count, not two.
3. On November 2, 2015, a Major PDO AV crash involved three vehicles: one AV and two CVs. Teoh and Kidd (2017) [26] added only one crash to their AV crash count, not three.

In the AV-exposed group, there were nine crash-involved vehicles (4+2+3) in three AV crashes. But Teoh and Kidd (2017) [26] counted only the three AV crashes, not the nine vehicles involved in an AV crash. These AV crash examples prove that Teoh and Kidd (2017) [26] counted AV crashes at the crash level, while Table E.1 proves that Teoh and Kidd (2017) [26] counted CV crashes at the vehicle level. Using different criteria for selecting membership in the AV and CV groups (crashes for the AV group, but vehicles for the CV group) is a selection bias. This unique differential selection method by Teoh and Kidd (2017) [26] upwardly biased their CV crash count and upwardly biased their CV crash rate. These upward biases then downwardly biased the AV crash IRR (AVs would falsely appear more protective than they are). This selection bias largely explains why the AV IRR estimate of 0.36 in Table 5.1 (Chapter 5, Section 5.1.1.5), based on Teoh and Kidd's (2017) [26] original Phase I data, is substantially below one, the Mountain View CV comparison benchmark.

One method to adjust for the crash selection bias is to use the same "vehicle-involved" count criterion for both AV crashes and CV crashes. If Teoh and Kidd (2017) [26] had done that, there would have been nine police-reportable AV crashes instead of three in Table 5.1 (note b). (Although only three vehicles were AVs, a total of nine vehicles were crash-involved with the AV crashes.) Therefore, to be consistent in their AV and CV crash selection criteria, Teoh and Kidd (2017) [26] should have used nine vehicles involved in AV crashes in Table 5.1 (note b) instead of three AV crashes. Using Teoh and Kidd's (2017) [26] data in Table 5.1 for the other three estimates in the 2×2 table (AV mileage and Mountain View CV crash-involved vehicles and mileage), but with nine instead of three AV crashes, the Google AV IRR would increase from 0.36 to 1.08 (95% CI 0.53–2.0), exact two-sided $P = 0.77$. In short, there would then be only a trivial difference between the AV and CV crash-involvement rates, unlike the substantial AV reduction in crash rate compared to the CV crash rate as claimed by Teoh and Kidd (2017) [26].

Another way to adjust for the crash selection bias is to use the number of CV crashes (4,862) in Table 5.1 (note f) to replace the number of CVs "involved" in crashes. That way, both the AV group and CV group count crashes, so "apples are being compared to apples," which is the usual AV-to-CV crash rate comparison method. That method gives rise to an AV crash IRR estimate of 2.11 (95% CI 0.55–5.8), with an exact two-sided $P = 0.23$ (Section 5.1.3, Table 5.2). That method is the standard way all the other final AV crash IRRs were estimated in the reanalyses in this review.

In summary, the "vehicle-involved" CV selection method used by Teoh and Kidd (2017) [26] contained a substantial crash selection bias because their method used different selection criteria for the AV group versus the CV group. This crash selection bias in the Teoh and Kidd (2017) [26] analysis made AVs seem safer than they are when compared to passenger vehicles.

One way to remove the crash selection bias is to count crash-involved vehicles in the AV group, matching the criterion of counting crash-involved vehicles in the CV group. There is then little difference in the AV and CV crash rates (IRR = 1.08).

A second way to remove the crash selection bias is to count only crashes in both the AV and CV groups (not the total vehicles involved in the crashes). But the result is a somewhat elevated AV IRR = 2.11.

Teoh and Kidd (2017) [26] give no justification as to why a vehicle-level analysis is better than a crash-level analysis. If a vehicle-level analysis is used, it must have the same criteria for the AV and CV groups, unlike what Teoh and Kidd (2017) [26] did. However, none of the other five reviewed studies in this book used a vehicle-level analysis when comparing AVs to CVs. Therefore, vehicle-level analyses were not pursued further. In reanalyzing Teoh and Kidd's (2017) [26] original data, a crash-level analysis was used for the AV and CV groups (see Section 5.1.3, Table 5.2).

Appendix J
Crash Reduction: AVs versus Crash Avoidance Technologies

Note: Refer to Appendix B on page 149 for an explanation of the abbreviations and the definitions of unique terms and phrases.

The "Limitations" Section 9.5 in the Discussion in Chapter 9 raised the question of how the police-recorded Google AV crash reductions found in Table 2.5A (Chapter 2, Section 2.4.1) compare to the crash reductions from crash avoidance technologies (Yue et al., 2018) [55]). Specifically, do AVs reduce crashes more than the crash avoidance technologies already in the marketplace? If yes, then AVs offer improved traffic safety. If not, then improved traffic safety might be more cheaply and readily accomplished by increasing the penetration of crash avoidance technologies in the marketplace rather than converting the U.S. fleet to AVs.

Yue et al. (2018, Table 4a) [55] estimate that connected vehicle and driver assistance technologies reduce police-recorded crashes of human-driven light vehicles by a conservative 33% reduction to an aggressive 46% reduction. These crash reductions by crash avoidance technologies seem comparable at first to the observed 43% reduction in the Severe AV crash rate in the standardized Google AV data (Table 2.5A). This comparison suggests that AVs offer little or no safety benefit over crash avoidance technologies. However, this comparison overlooks two substantial biases.

First, Yue et al. (2018) [55] found a crash reduction effect from crash avoidance technologies only for police-recorded crashes. Their article did not indicate to what extent crash avoidance technologies reduce all crashes, including those not recorded by police. However, the standardized Google AV Severe crash rate reductions (Table 3.2A) were for all AV Severe crashes, police-recorded and not police-recorded. This unrecorded crash selection bias is present when directly comparing crash reductions of AVs versus crash avoidance technologies.

Second, the Google AV 43% crash reduction in Table 3.2A was for Severe crashes only. Yue et al.'s (2018) [55] crash reductions from crash avoidance technology were for police-recorded crashes at all severity levels (fatal, injury, and PDO). However, Yue et al. (2018) [55] did not provide crash data stratified by crash severity level. Hence, a reduction specifically in the severe crash rate is not available from Yue et al.'s (2018) [55] analysis. So a direct comparison of the Yue et al. (2018) [55] total crash reduction from crash avoidance technologies to the Severe crash reduction from the Google AV fleet (Table 3.2A, Section 3.4.1, Chapter 3) has a second crash selection bias because of the different crash severity levels.

These two biases can be avoided using the AV data in Table 2.5 (Section 2.4.1, Chapter 2) instead of the AV data in Table 3.2 (Section 3.4.1, Chapter 3). Table 2.5 is only police-recorded data, so it is not subject to crash selection bias, as is Table 3.2 (which shows all crashes, not just police-recorded crashes). Therefore, a valid comparison can be made of the AV crash reductions in Table 2.5 to the results of Yue et al. (2018) [55]. Table 2.5 indicates a 100% reduction in police-recorded AV crashes at every crash severity. Therefore, the severity strata were homogeneous, so they could be combined without violating the heterogeneity best

TABLE J.1 Google AV police-recorded crash rate versus three levels of total crash reduction in 2013 U.S. CVs from crash avoidance technologies.

	A. No reduction		B. Conservative 33% reduction		C. Aggressive 46% reduction	
	Google AV	2013 U.S. CV	Google AV	2013 U.S. CV	Google AV	2013 U.S. CV
Total police-recorded crashes	**0**	**5,687,057**	**0**	**3,810,897**	**0**	**3,095,103**
MVMT	**1.0437**	**2,988,323**	**1.0437**	**2,988,323**	**1.0437**	**2,988,323**
Crash rate	0.00	1.89	0.00	1.28	0.00	1.04
IRR (95% CI)	0.0 (0–1.5), $P = 0.14$		0.0 (0–2.3), $P = 0.26$		0.0 (0–2.7), $P = 0.34$	

Table A is the same as Table 2.5D (Section 2.4.1, Chapter 2). Table B shows a 33% "conservative" reduction in CV crashes, and Table C an "aggressive" 46% reduction in CV crashes if crash avoidance technologies had been in place in 2013.

A **bold** number in a 2×2 subtable means no identified error or bias.

practice. The total AV crash rate in Table 2.5D can then be compared to the total crash rate reductions investigated by Yue et al. (2018) [55], avoiding a crash severity bias.

In particular, the data in Table 2.5D (Section 2.4.1, Chapter 2) in the main body of this book estimates a Google AV known police-recorded total crash IRR of zero. This zero result indicates that Google AVs decrease the total police-recorded crash rate by 100% versus the corresponding 2013 U.S. police-recorded CV crash rate. But with zero AV crashes, there would be a zero police-recorded AV crash rate and a zero IRR compared to any CV benchmark.

Therefore, the total AV crash reduction data in Table 2.5D can be directly compared to the Yue et al. (2018) [55] total crash reduction data for crash avoidance technologies. In other words, one can validly use the police-recorded Google AV crash reductions in Table 2.5D to compare to the crash avoidance technology reductions in Yue et al. (2018) [55].

Table J.1 compares the conservative and aggressive crash reductions by crash avoidance technologies as estimated by Yue et al. (2018, Table 4a) [55] to the Google AV total police-recorded crash rate of zero from Table 2.5D.

Tables J.1A–C (left columns) show the Google AV police-recorded crash rate of zero estimated in Table 2.5 (Section 2.4.1). Tables J.1B and J.1C show that Google AVs decrease the total police-recorded crash rates more than the "conservative" or "aggressive" reductions in police-recorded crashes from crash avoidance technologies.

Nonetheless, if valid, these data are also reasonably compatible with the Google AV total police-recorded crash rate ranging from a 100% decrease in Tables J.1A–C at the lower confidence limit (0) of the Google AV crash IRR estimates. Table J.1A estimates a 51% increase from AVs at the upper confidence limits of the IRR estimates compared to CVs with no crash avoidance technology. Table J.1B estimates a 125% AV crash increase compared to a conservative crash avoidance technology reduction. Finally, Table J.1C estimates a 173% AV crash increase versus an aggressive crash avoidance technology reduction.

In conclusion, the crash rate comparisons in Table J.1 indicate that Google AVs reduce the total crash rate relative to crash avoidance technologies. However, the data are not yet precise enough to make that claim compared to the upper confidence limits of the Google AV total police-recorded crash rate estimate, so the data are not yet definitive.

The data in Table J.1 are not definitive for another reason. Why are there no matches in the CA SWITRS [28] police database records to any Google AV crashes? That fact is despite there being Severity Level 1 Google AV crashes with injuries and Severity Level 2 (Major PDO) Google AV crashes that exceed the $750 police reporting requirements in CA at the time of the AV field tests (see Appendix E). Therefore, it is plausible that there are still biases contributing to the lack of police-reported AV crashes in these early Google AV crashes in the six studies critically reviewed here.

TABLE J.2 Mountain View site bias.

	2013 Mountain View CV[a]	2013 U.S. CV[e]
Total police-recorded* crashes	**178**[b]	**5,687,057**[f]
MVMT[i]	**205.70**[c]	**2,988,323**[g]
CV total crash rate	0.865[d]	1.903[h]
IRR (95% CI)	0.45 (0.39–0.53), $P < 0.0000001$	

2013 total police-recorded CV crash rate: Mountain View versus U.S. national.

A **bold** number in the 2×2 table means no identified error or bias.

Notes:

[a] Mountain View CV data from January 1, 2013, to December 31, 2013.

[b] Total (fatal, injury, and PDO) police-recorded crashes on Mountain View city streets (SWITRS, 2013) [28].

[c] Millions of miles traveled on Mountain View city streets (HPMS Report, 2013) [24].

[d] CV crash rate in Mountain View city streets.

[e] 2013 U.S. national CV data (NHTSA, 2015, p. 2) [56] (same as Tables 2.1 and 2.5D in the main body).

[f] Police-recorded fatal + injury + PDO crashes.

[g] Estimated MVMT on all U.S. roads.

[h] 2013 U.S. CV police-recorded total crash rate estimate.

[i] Millions of Vehicle Miles Traveled.

For example, a site selection bias contributed to the zero police-recorded Google AV crashes in the Mountain View area. All the Google AV crashes took place in Mountain View and vicinity. Table J.2 compares the 2013 Mountain View city street CV police-recorded crash rate to the 2013 U.S. CV police-recorded crash rate in Table 2.5D and Table J.1A.

The data in Table J.2 estimate a Mountain View city street police-recorded total crash IRR of 0.45. This result indicates that the Mountain View city street site decreases the police-recorded total CV crash rate by 55% versus the 2013 U.S. national police-recorded CV total crash rate. Nonetheless, these data, if valid, are also reasonably compatible with this police-recorded total CV crash rate at the Mountain View city street site ranging from a decrease of 47% to 61% versus the 2013 U.S. national police-recorded total CV crash rate. (This 2013 result in Table J.2 is consistent with the Table K.1 result for the same comparison but using the 2015 Mountain View CV versus the 2015 U.S. national CV crash rate estimate.)

This lower overall police-recorded crash rate in Mountain View partly explains why Mountain View has no police records for any of the 11 Google AV crashes (Table 2.1, note b, Section 2.1.1, Chapter 2) in the Schoettle and Sivak (2015) [31] study. Also, there is a 55% lower rate of total police-recorded CV crashes in Mountain View than in the United States nationally (Table J.2). However, it is not yet determined if (1) there is simply a lower CV crash rate in Mountain view, (2) drivers in Mountain View report crashes at a lower average rate than nationally (see Appendix E, Section E.1), or (3) the police threshold for recording a crash is higher in Mountain View than the U.S. average (see Appendix E, Section E.2).

However, there can also be more specific reasons why AVs on city streets in Mountain View and vicinity have zero police-recorded crashes per Table 2.5 and Table J.1. A discussion of these more specific reasons can be seen in Appendix E, Section E.5.

The bottom line is that no definitive data yet exists by which one can compare crash reduction effects from AVs to crash reduction effects in CVs from crash avoidance technologies. The lack of an adequate CV control group is the main issue. This issue can be readily solved by requiring AV companies to release the crashes and mileage for three conditions:

1. The AVs in autonomous mode in their ODD (already reported to the CA DMV).
2. The same AVs in the same ODD but driven by a human driver (manual mode), with crash avoidance technologies providing warnings to the driver (not yet released).
3. The same AVs in the same ODD but driven by a human driver (manual mode), with no crash avoidance technologies providing warnings to the driver (not yet released).

Appendix K
Site Selection Bias: Site Differences in CV Crash Rates

CONTENTS

Note: Refer to Appendix B on page 149 for an explanation of the abbreviations and the definitions of unique terms and phrases.

Appendix K investigates site selection bias arising from site differences in CV crash rate estimates. It shows crash rate differences between Mountain View, San Francisco, and the U.S. crash rate nationally. Road selection bias and environmental selection bias are also mentioned.

K.1. Mountain View versus U.S. CV Total Crash Rate

Table K.1 compares the 2015 Mountain View crash rate estimate to the preliminary 2015 U.S. national crash rate estimate.

The data in Table K.2 estimate an IRR of 0.72. In other words, in 2015, the police-recorded CV total crash rate in Mountain View was about 28% lower than in the United States as a whole. (This IRR estimate below one is consistent with the IRR estimate below one in Table J.2 for the same comparison but using 2013 data.)

Does that mean that Mountain View has safer drivers? Not necessarily, because the reduced crash rate in Mountain View still has site selection biases. First, there is a road selection bias because the Mountain View police-recorded crash rate is for city streets only; the U.S. national crash rate includes all roads. In the SHRP2 study (Blanco et al., 2016, Figures 7-9) [27], interstates and divided highways had lower "overall" crash rates (PR and non-PR) than streets in moderate residential sites, for all non-fatal crash Severity Levels. In other words, driving in a "moderate residential" area like Mountain View has a higher overall crash rate than driving at many sites included in the national average, such as bypass/divided highway, open residential, open country, and interstate, at all non-fatal Severity Levels (Blanco et al., 2016, pp. 27–28, Figures 7–9) [27]; see also Chapter 2, Section 2.1.3). Blanco et al. (2016, Figures 7–9) [27] also showed that these road selection bias effects were consistent across crash Severity Levels. This pattern is less evident if only PR crashes are considered, for Level 1

TABLE K.1 Comparison of 2015 CV total police-recorded crash rate in Mountain View versus the preliminary 2015 U.S national.

	2015 Mountain View CV[a]	2015 U.S. CV[e]
Total police-recorded crashes	323[b]	6,296,000[f]
MVMT[i]	224.268[c]	3,148,000[g]
CV total crash rate	1.44[d]	2.00[h]
IRR (95% CI)[i]	0.72 (0.64-0.80), P = 1.5E-9	

Notes:

[a] 2015 human-driven CV data in Mountain View, CA.

[b] Total police-recorded CV crashes on city streets in Mountain View (SWITRS, 2015) [28].

[c] CV mileage on city streets in Mountain View in 2015 (HPMS, 2015) [24].

[d] CV police-recorded crash rate on Mountain View city streets.

[e] Preliminary 2015 U.S. national CV data.

[f] Total CV police-recorded crashes (NHTSA, 2016, Table 2) [57], Favarò et al. (2017, Table 4) [33].

[g] Total estimated 2015 CV mileage on all U.S. roads (city streets, state highways, and interstates) (Favarò et al., 2017, Table 4) [33]; (FHWA, 2015, p. 1) [58].

[h] Preliminary U.S. CV crash rate.

[i] IRR and exact mid-P confidence levels and two-tailed P estimated with StatsDirect rate ratio program [59].

(Severe) and Level 2 (Major PDO) crashes (Blanco et al., 2016, Figures 7 and 8) [27] because the numbers of PR crashes are too small to obtain reliable crash rate estimates.

Driving in a "moderate residential" area also has a lower CV crash rate than driving in an urban setting at all Severity Levels (Blanco et al., 2016, Figures 7-9) [27]. Therefore, the Mountain View IRR estimate is potentially biased relative to the U.S. crash rate nationally because of road selection bias. The bias direction would depend upon the relative exposure (mileage) in each of the sites in the national average. Another way of illustrating road selection bias is to restrict the U.S. national crash rate estimate to only "moderate residential" city streets. Then the Mountain View crash IRR estimate would likely change from 0.72 for reasons other than Mountain View drivers.

A second bias in the Mountain View IRR in Table K.2 is environment selection bias. Environmental conditions are milder in Mountain View than the U.S. average, giving rise

TABLE K.2 Comparison of 2015 CV total police-recorded crash rate in Mountain View versus San Francisco.

	2015 Mountain View CV[a]	2015 San Francisco CV[e]
Total police-recorded crashes	323[b]	4,562[f]
MVMT	224.268[c]	1,952.146[g]
Crash rate	1.44[d]	2.34[h]
IRR (95% CI)[i]	0.62 (0.55-0.69), P = 2.0E-11	

Notes:

[a-d] Same as Table K.1.

[e] 2015 CV data on city streets in San Francisco.

[f] Total police-recorded CV crashes on city streets in San Francisco (SWITRS, 2015) [28].

[g] 2015 CV mileage on city streets in San Francisco (HPMS, 2015) [24].

[h] 2015 San Francisco CV crash rate.

[i] Same as Table K.1.

to comparatively lower crash rates in Mountain View. If the U.S. national crash rate could be estimated only for regions with mild environmental conditions, the Mountain View versus national IRR estimate would likely be higher than 0.72, again for reasons having nothing to do with Mountain View CV drivers.

Schoettle and Sivak (2015) [31] were the first to recognize that these biases would likely affect AV fleet crash rate estimates (see Chapter 2, Sections 2.1.3 and 2.4.2). Those investigators hypothesized that the same risk factors that affect the CV crash rate would also likely affect the AV crash rate. That hypothesis is confirmed by evidence that the AV crash rate trend closely follows the CV crash rate trend at the same site (compare Figure 2.2A to Figure 2.2B and Figure 5.2A to Figure 5.2B).

In summary, Table K.1 indicates that the crash rate of an AV fleet with its ODD primarily in Mountain View (such as Google/Waymo) cannot be validly compared to the U.S. national CV crash rate. It must have an adjustment for road selection bias and site selection bias. A valid comparison of an AV to any CV crash rate thus requires adjusting CV crash rate differences between the ODD site and the CV site. The usual epidemiological adjustment method for site differences is termed standardization (see Section 3.4 for further discussion and Appendix H for a numerical example).

K.2. Mountain View versus San Francisco CV Total Crash Rate

Table K.2 compares the 2015 Mountain View CV crash rate estimate to the 2015 San Francisco CV crash rate estimate.

The data in Table K.2 estimate an IRR of 0.62, indicating a 38% lower police-recorded CV total crash rate in Mountain View than San Francisco. What are the possible reasons for this lower police-recorded CV crash rate?

1. Note that the IRR estimate in Table K.2 has a different type of site selection bias than does Table K.1. The Mountain View and San Francisco crash rate estimates in Table K.2 are for city streets only, and both sites have relatively mild Bay Area environmental conditions. However, the "city" streets in Mountain View are in a predominately "moderate residential" site. The city streets in San Francisco are in an urban site. Driving in a moderate residential site is known to have a lower crash rate than driving in an urban site at all non-fatal crash Severity Levels (Blanco et al., 2016, Figures 7–9) [27]. Thus an AV test with an ODD in Mountain View would be biased towards a lower crash rate than an AV test with an ODD in San Francisco, even if it were the identical AV fleet at the two sites.
2. A second possibility for the relatively low crash rate for the Mountain View site is that Mountain View drivers involved in a crash might have a lower probability of reporting the crash to the police than San Francisco drivers.
3. Given that a crash has been reported to the police, there is a third possible explanation. The Mountain View Police Department might not convert a crash bench report to a Police Accident Report as often as the San Francisco Police Department.
4. Fourth, if the Mountain View Police Department wrote up a Police Accident Report, they still might not choose to file an actual crash record in the SWITRS state police database [28] as frequently as does the San Francisco Police Department.

See Appendix E for further discussion of the biasing effects of police reporting and police recording practices.

K.3. San Francisco versus U.S. CV Total Crash Rate

Table K.3 compares the 2015 Mountain View CV crash rate estimate to the 2015 San Francisco CV crash rate estimate.

The data in Table K.3 estimate an IRR of 1.17, indicating a 17% higher police-recorded CV total crash rate in San Francisco compared to the United States. This result is not surprising because driving in an urban site has the highest CV crash rate of any site at all Severity Levels. Blanco et al. (2016, p. 27) [27] state, "For all three crash [Severity] types, the highest crash rates occurred in Urban areas: 6.6 per million miles (age-adjusted) for Level 1, 8.9 per million miles of driving for Level 2 (age-adjusted), and 53.7 per million miles for Level 3 (age-adjusted)."

However, these high rates for urban areas in the SHRP2 study sites were for total crashes, PR and not PR, and across all crash severities. For only PR age-adjusted SHRP2 Severe crashes, urban areas had a more mixed crash rate relationship to other sites. Figure 7 in Blanco et al. (2016) [27] shows an approximately equal Severe crash rate for urban areas to "moderate residential" and interstate areas. But urban areas had a lower Severe crash rate than business-industrial, schools, bypass/divided highway, "open residential," and "open country" areas. However, the small numbers of PR Severe crashes make these comparisons uncertain. But for Major PDO crashes, the Blanco et al. (2016, Figure 8) [27] data clearly show that urban areas had a higher PR crash rate than all other areas.

The IRR estimate of 1.17 in Table K.3 has additional biasing factors, as discussed in Section K.2. For example, the San Francisco police-recorded crash rate is for city streets only, but the U.S. national crash rate includes all roads (road selection bias). In addition, the environmental conditions are different in San Francisco than in the United States as a whole (environment selection bias).

As a limitation of these analyses, note that they are for the total crash rate across all crash severities. Because of these site differences, any differences in AV crash rates with an ODD in San Francisco cannot be validly compared to the U.S. national CV crash rate without some method of adjustment for these and other biases. Again, the usual epidemiological adjustment method to adjust for site differences is termed standardization (see Section 3.4 and Appendix H).

TABLE K.3 Comparison of 2015 CV total police-recorded crash rate in San Francisco versus U.S. national.

	2015 San Francisco CV[a]	2015 U.S. CV[e]
Total police-recorded crashes	4,562[b]	6,296,000[f]
MVMT	1,952.146[c]	3,148,000[g]
Crash rate	2.34[d]	2.00[h]
IRR (95% CI)[i]	1.17 (1.12–1.21), $P = 1.6E-15$	

Notes:

[a-d] Same as Table K.2, notes e–h.

[e-i] Same as Table K.1, notes e–i.

Appendix L
Summary of Issues

Note: Refer to Appendix B on page 149 for an explanation of the abbreviations and the definitions of unique terms and phrases.

Appendix L tabulates the main results for the six studies reviewed in Chapters 2, 3, 4, 5, 6, and 7. Section 8.1 in Chapter 8 gives an overview of this Appendix. Table L.1 is a summary of the complete Table L.2.

The rows of Table L.1 are the original and final AV IRR estimates for each study, which are summarized for each study in Section 8.2, Chapter 8. The final AV IRR estimates are plotted in Figure 8.1 and discussed in Section 9.1, Chapter 9.

The columns of Table L.1 are the 11 issues for the six studies. Each issue is summarized in Section 8.3. The final issues are also discussed in Section 9.2.

Table L.2 gives the exact section, table, and figure number for the data underlying each of the entries in Table L.1. Table L.2A tabulates the systematic errors and selection biases (columns) for each study (rows). Table L.2B summarizes the selection biases across sites and best practice violations of heterogeneity(columns) for each study (rows).

After completing the manuscript, a new bias was discovered in the standardization method in Appendix H that was used for the Blanco et al. (2016) [28] study data. This bias is mentioned at the end of Appendix H and discussed in Appendix N. Thus, the final Blanco et al. (2016) [28] IRR data in Table L.1 is not 100% free of bias as indicated in Table L.1. A crash selection bias remains in the standardized Google AV crash rates in Table 3.2 and Figure 3.5 and the standardized Google AV crash IRR estimates in Table 3.2 and Figure 3.6. See Appendix N for details.

TABLE L.1 Summary of Table L.2. Systematic errors, selection biases, and best practice violations in the six reviewed original studies and after final reanalysis.

Chapter & study		Systematic errors		Selection biases: Crash and mileage				Selection biases: Sites		Best practice violations: Heterogeneity			Test result			
		1. Minor	**2. Substantial**	**3. AV Crashes**	**4. AV MVMT**	**5. Unrecorded CV Crashes**	**6. CV MVMT**	**7. CV crash rate diff., AV vs. CV site**	**8. Non-CA AV test sites**	**9. AV companies**	**10. Crash severity**	**11. CV sites**	**Y**	**N + ~A[1]**	**%Y**	**Result**
2. Schoettle and Sivak (2015) [31]	**Orig.**	**N**	N/A	N	N	**N**	Y	**N**	**N**	N	N	N/A	1	8	11%	Not Meet
	Final	**Y**	N/A	**Y**	**Y**	**Y**	**Y**	**N**	Y	Y	Y	N/A	8	1	89%	Meet[2]
3. Blanco et al. (2016) [27]	**Orig.**	**Y**	N/A	**Y**	**Y**	**Y**	**Y**	**N**	**N**	N/A	Y	**N**	6	3	67%	Not Meet
	Final	**Y**	**Y**	**Y**	**Y**	**Y**	**Y**	**Y**	**Y**	N/A	Y	**Y**	10	0	100%	Meet[2]
4. Dixit et al. (2016) [13]	**Orig.**	**Y**	**N**	**Y**	**Y**	**N**	**N**	**N**	N/A	N/A	N	N/A	3	5	38%	Not Meet
	Final	**Y**	**Y**	**Y**	**Y**	**N**	**Y**	**N**	N/A	N/A	~A	N/A	5	3	63%	Not Meet
5.Teoh and Kidd (2017) [26]	**Orig.**	**N**	**N**	**N**	**N**	**N**	**N**	**N**	N	N/A	N	N	0	10	0%	Not Meet
	Final	**Y**	**Y**	**N**	**Y**	**Y**	**Y**	**Y**	Y	N/A	N	Y	8	2	80%	Meet[2]
6. Favarò et al. (2017) [33]	**Orig.**	N	N/A	Y	Y	**N**	Y	~A	Y	N	N	N	4	6	40%	Not Meet
	Final	Y	N/A	Y	Y	~A	Y	~A	Y	~A	~A	~A	5	5	50%	Not Meet
7. Banerjee et al. (2018) [22]	**Orig.**	N	**N**	Y	Y	**N**	Y	**N**	Y	N/A	~A	~A	4	6	40%	Not Meet
	Final	**Y**	~A	Y	Y	~A	Y	~A	Y	N/A	~A	~A	5	5	50%	Not Meet
Count OK (Y)	**Orig.**	2	0	4	4	1	4	0	2	0	1	0				
	Final	6	3	5	6	3	6	2	5	1	2	0				
Not OK (N + ~A)[1]	**Orig.**	4	3	2	2	5	2	6	3	2	5	4				
	Final	0	1	1	0	3	0	4	0	1	4	2				
%OK ÷ Total Appl.[3]	**Orig.**	33%	0%	67%	67%	17%	67%	0%	40%	0%	17%	0%				
	Final	100%	75%	83%	100%	50%	100%	33%	100%	50%	33%	50%				

Key:

Green Y indicates "OK." There is no identified error, bias, or best practice violation for this study for this metric.

Red N indicates "Not OK." There is an identified error, bias, or best practice violation for this study for this metric.

Bold Y or **N** indicates the issue has a section devoted to it in the main body (see Table L.2 for cross-reference).

~A = Not Addressed.

N/A = Not Applicable.

Notes:

[1] "N + ~A" = count of Not OK (N) + Not Addressed (~A)

[2] Studies with the top three final %Y test results after corrections and adjustments. These were the only studies that met a 75% threshold for the percentage of tests free of the 11 issues in the final reanalysis. The final IRRs from the three studies that met this final validity threshold are shown in Figure 8.2 in the main body.

[3] The "Total Applicable" count is (Y + N + ~A). This count excludes the "Not Applicable" (N/A) cells.

TABLE L.2A Tabulation of systematic errors and selection biases in crashes and mileage in the six original studies and final reanalysis.

Chapter/ article		Systematic errors		Selection biases: Crash and mileage			
		1. Minor	2. Substantial	3. AV crashes	4. AV MVMT	5. Unrecorded CV crashes	6. CV MVMT
2. Schoettle and Sivak (2015) [31]	Orig.	**§2.2.1,**T2.1-2,4f	N/A	§2.1.1,T2.1b	§2.1.1,T2.1c	**§2.1.1-2,**T2.1f	§2.1.1,T2.1g
	Final	**§2.3**,T2.5f	N/A	**§2.4.1**,T2.5b	**§2.4.1**,T2.5c	**§2.4.1**,T2.5f	**§2.4.1**,T2.5g
3. Blanco et al. (2016) [27]	Orig.	**§3.1**,T3.1,F3.1	N/A	**§3.1**,T3.1b,F3.1	**§3.1**,T3.1c,F3.1	**§3.1**,T3.1f,F3.1	**§3.1**,T3.1g,F3.1
	Final	**§3.4.1**,T3.2,F3.5	**§3.4.1**,T3.2,F3.5	**§3.4.1**,T3.2b,F3.5	**§3.4.1**,T3.2c,F3.5	**§3.4.1**,T3.2f,F3.5	**§3.4.1**,T3.2g,F3.5
4. Dixit et al. (2016) [13]	Orig.	**§4.2.3**,T4.1g	**§4.2.1**,T4.1f	**§4.2**,T4.1b	**§4.2**,T4.1c	**§4.2.2**,T4.1f	**§4.2.3**,T4.1g
	Final	**§4.2.4**,T4.2g	**§4.2.4**,T4.2f	**§4.2.4**,T4.2b	**§4.2.4**,T4.2c	**§4.2.5**,T4.2f	**§4.2.4**,T4.2g
5.Teoh and Kidd (2017) [26]	Orig.	**§5.1.2.8**,T5.1g	**§5.1.2.5-7**,T5.1f	**§5.1.2.1-2**,T5.1b	**§5.1.2.1**,T5.1c	**§5.1.2.5-7**,T5.1f	**§5.1.2.9-10**,T5.1g
	Final	**§5.1.3**,T5.2g	**§5.1.3**,T5.2f	**§5.1.3**,T5.2b	**§5.1.3**,T5.2c	**§5.1.3**,T5.2f	**§5.1.3**,T5.2g
6. Favarò et al. (2017) [33]	Orig.	§6.1,T6.1gh	N/A	§6.1,T6.2b	§6.1,T6.2c	**§6.2.3**,T6.2f	§6.1,T6.2g
	Final	§6.1,T6.2gh	N/A	§6.1,T6.2b	§6.1,T6.2c	Not Addr.	§6.1,T6.2g
7. Banerjee et al. (2018) [22]	Orig.	§7.1,T7.1k	**§7.2.2**,T7.1dg	§7.1,T7.1b	§7.1,T7.1c	**§7.2.4**,T7.1fghi	§7.1,T7.1c
	Final	**§7.2.1**,T7.1k	Not Addr.	§7.1,T7.1b	§7.1,T7.1c	Not Addr.	§7.1,T7.1c

TABLE L.2B Summary of selection biases across sites and best practice violations of heterogeneity in the six original studies and after final reanalysis.

Chapter/article		Selection biases: Sites		Best practice violations: Heterogeneity		
		7. CV crash rate diff., AV vs. CV site	**8. Non-CA AV test sites**	**9. AV companies**	**10. Crash severity**	**11. CV sites**
2. Schoettle and Sivak (2015) [31]	**Orig.**	**§2.1.3,**T2.2d	**§2.2.2,**T2.2c	**§2.2.3,**T2.2d	**§2.2.4,**T2.2bf	N/A
	Final	**§2.4.2**	§2.3,T2.4c	§2.3,T2.4d	§2.3,T2.4bf	N/A
3. Blanco et al. (2016) [27]	**Orig.**	**§3.3,**T3.1d,F3.1-2,F3.3-4	**§3.3,**T3.1c,F3.1-2,F3.3-4	N/A	**§3.1,**T3.1bf,F3.1	**§3.3,**T3.1c,F3.1-2,F3.3-4
	Final	**§3.4.1,**T3.2d,F3.5-6	**§3.4.1,**T3.2c,F3.5-6	N/A	**§3.4.1,**T3.2bf,F3.5-6	**§3.4.1,**T3.2c,F3.5-6
4. Dixit et al. (2016) [13]	**Orig.**	**§4.2.6,**T4.1d	N/A	N/A	T4.1bf	N/A
	Final	**§4.2.6,**T4.2d	N/A	N/A	Not Addr.	N/A
5. Teoh and Kidd (2017) [26]	**Orig.**	**§5.1.2.3-4,**T5.1d	§5.1.1.5,T5.1a	N/A	**§5.1.2.1-2,**T5.1-2bf	§5.1.1.5,T5.1a
	Final	**§5.1.3,**T5.2d	§5.1.3,T5.2a	N/A	**§5.1.3,**T5.2bf	§5.1.3,T5.2a
6. Favarò et al. (2017) [33]	**Orig.**	Not Addr.,T6.2d	§6.1,T6.2c	**§6.2.1,**T6.1-2d	6.1-2bf	**§6.2.2,**T6.1-2d
	Final	Not Addr.	§6.1,T6.2c	Not Addr.	Not Addr.	Not Addr.
7. Banerjee et al. (2018) [22]	**Orig.**	**§7.2.5,**T7.1dghi	§7.1,T7.1bc	N/A	Not Addr.	Not Addr.
	Final	Not Addr.	§7.1,T7.1bc	N/A	Not Addr.	Not Addr.

Key to Tables L.2A and L.2B:

Red indicates "Not OK." There is an identified error, bias, or best practice violation in the data analysis.

Green indicates "OK." There is no identified error, bias, or best practice violation in the data analysis.

"§" indicates the Section number in the indicated chapter that is relevant to the issue.

Bold green or **red** Section numbers indicate the Section describing that specific issue for a given study.

"T" indicates the Table number in the main body.

The letter(s) following a table number indicates the superscript of the table footnote that is affected by the issue.

"F" indicates the Figure number in the main body.

N/A = Not applicable.

Not Addr. = Not Addressed.

Appendix M
Do ADS Neural Networks Have Negative Learning?

CONTENTS

Note: Refer to Appendix B on page 149 for an explanation of the abbreviations and the definitions of unique terms and phrases.

An upward trend in the Google AV crash rate was noted in Figure 2.2A (Section 2.6.2.1, Chapter 2) and Figure 5.2A (Section 5.3, Chapter 5).

Section 2.4.2.2.2 advanced the hypothesis that this upward crash rate trend may have resulted from negative learning by the neural network underlying an ADS. Google [30] has stated that it frequently updates its onboard ADS software by putting the on-road test results through its simulator and trying multiple crash scenarios with variations. The "improved" software is then downloaded to the vehicle with the expected outcome of reducing crashes and disengagements by the test driver and ADS. But an ADS neural network might have a negative learning effect, where more training might cause more driving performance errors, not fewer. The situation might vary depending upon whether the ADS neural network is equivalent to (1) an "overdetermined" system of mathematical equations, (2) an "underdetermined" system, or (3) neither.

M.1. "Overdetermined" ADS?

An "overdetermined" system arises if an ADS neural network has more training examples than parameters. (Although "parameters" are a mathematical construct and are not artificial neurons, the weights and biases in a neural network are here considered to have similar mathematical properties to "parameters.") Such a neural network might perform like an overdetermined system of equations in mathematics. When the equations have some underlying structure, such as that which occurs with driving, then the neural network system might have a unique solution or an infinite number of solutions. However, in practice, and depending on the application, an overdetermined system can yield useful approximate solutions by some method like least-squares approximation (finding a point whose perpendicular distance from each equation is minimized). An overdetermined neural network might thus be better conceptualized as a least-squares method than an exact solution method, more like statistics than algebra.

For driving applications, that means an overdetermined neural network might make driving response errors (i.e., crashes) on the training examples in the driving simulator, no matter how many training examples it was given. Even more errors might occur in real-world driving because the errors will occur not only for objects and events on which the ADS has been trained but also for new objects and events the ADS encounters on the road.

Thus, additional training examples for an overdetermined ADS might not guarantee that AV crash rates would improve with practice. Theoretically, crashes might even increase with increased training of an overdetermined ADS. However, many experts prefer an underfitted model with fewer parameters (an overdetermined system), which reasonably reproduces larger data sets. It is also considered more trustworthy than an overfitted model (see next section), which more closely reproduces the data set on which it was trained. Some of the most successful neural nets achieved their success mainly because they had few parameters and thus little room for overfitting, which increased their range of applicability beyond the original training set.

M.2. "Underdetermined" ADS?

The converse "underdetermined" system arises if the ADS neural network has <u>fewer</u> training examples than parameters. For example, consider a vast neural network with a huge number of artificial neurons (here considered parameters). In this situation, there will always be more parameters than training examples. Consequently, the network would perform like an <u>underdetermined</u> system of equations in mathematics, in which "overfitting" can occur. The outcomes can become unreliable with overfitting when the system encounters new objects or events not included in the training set.

Underdetermined algebraic systems have infinite solutions, but what that suggests in practice depends on the application. Of course, there is an unlimited number of possible new objects and events that can be encountered in driving, so they cannot all be captured with the simple logic of accounting for each. But there might be only a finite number of "crash cases" in the real-world driving domain that have practical significance, even though there is literally an unlimited number of novel physical configurations. A finite number of heuristic rules might even be enough to handle almost all situations that would be considered reasonable. So, for a vehicle, the heuristic rule "if a large unknown object or event suddenly appears on the road in front of you, stop" would be a single rule that might help prevent a crash in a large number of scenarios. The caveat is if the AV stops suddenly and unexpectedly in the middle of the road in which a sizeable unknown object suddenly appears. In such a case, the probability of a CV (or another AV) crashing into the AV rear end increases substantially. In addition, if the momentum of the AV is too great and the time too short, the engineering mechanics could make it physically impossible for the AV to stop in time to avoid a crash into the object.

In an underdetermined system, the neural network can eventually be trained to make no driving errors in the set of training examples given in the driving simulator. But an underdetermined network can be overfitted and, therefore, can be unreliable and still make errors in responding to new objects and events in real-world driving. The main reason is that an underdetermined network may not generalize correctly to objects and events on which it has not been trained. However, additional valid training examples for an underdetermined ADS will reduce its ADS error rate, unlike for an overdetermined ADS. But there is an unlimited number of new objects and events in the real world. It follows that there will always be a certain number of driving errors by an ADS using underdetermined neural network technology, no matter how many artificial neurons and how much training the network has. Recent neural network research focuses on sophisticated procedures to minimize such errors. Experts say there are now many methods to ensure that a trained neural network will not overfit. However, naïve users who use neural networks (and other maximum-likelihood methods that use training sets) can overfit their training data, so they may not successfully address cases not captured by the training set.

M.3. Neither Overdetermined nor Underdetermined ADS?

The claim here is that it is too high a level of abstraction to compare neural networks to overdetermined or underdetermined sets of equations. Hence, neither of the above two sections are convincing, not because the mathematics is wrong but because the mathematics is far less relevant than the validity of the driving model. In other words, the skill with which the real-world driving domain is modeled is the critical factor that no mathematical model, neural network method, or millions of miles of driving in a simulator or the real world can avoid. Neural networks are, conceptually speaking, at the same level as statistical methods such as Maximum Likelihood, with all the same factors that make such methods better or worse for real-world applications. So it does not matter much if a neural network is overdetermined or underdetermined. The critical factor is the quality of the model of the real-world ODD.

Appendix N
PR Crash Selection Bias

CONTENTS

Note: Refer to Appendix B on page 149 for an explanation of the abbreviations and the definitions of unique terms and phrases.

There is a limitation to the Standardization method described in Appendix H. The "Standardization factor" (Table H.1, Var. #3) is defined by the "SHRP2 6 sites CV PR crash rate" (Var. #1) divided by the "Mountain View CV police-recorded crash rate" (Var. #2). However, PR crashes and police-recorded crashes use different crash selection criteria, so there is a remaining PR crash selection bias in the standardized metrics.

N.1. PR Crashes

The PR crashes at the six SHRP2 sites were defined in the SHRP2 study by Blanco et al. (2016) [27] as crashes with known Police Accident Reports (PARs). There were 912 known total crashes in the SHRP2 study (see the last line of Table E.1). Only 46 of those crashes (5% of 912) had a PAR [27]. This fact was determined after an extensive investigation by the SHRP2 investigators [27] with the cooperation of the local police departments at the six SHRP2 sites. However, it was not specified in the SHRP2 study [27] how many of those PAR crashes were entered as police-recorded crashes into a state crash database by the local police departments. The VTTI investigators [27] did not investigate how many of the 46 PARs were matched to a crash in that crash database to make them police-recorded crashes.

N.2. Police-Recorded Crashes

The police-recorded CV crashes at the Mountain View site were defined by the count of the crash records in the SWITRS [28] state police database. It is unknown how many CV crashes on city streets in Mountain View had PARs that were not recorded in the state SWITRS database [28]. Thus, the police-recorded criterion for CV crashes in Mountain View (using crashes in the SWITRS [28] database) is a different crash selection criterion than PR crashes at the six SHRP2 sites. Indeed, there are always fewer police-recorded crashes than PAR crashes at any given site. That is because the de-identified information in many PARs never gets uploaded to state crash databases, thereby becoming a police-recorded crash (see Appendix E). Hence, PR crashes are upwardly biased compared to police-recorded crashes. As a result, PR crash rates are also upwardly biased compared to police-recorded crash rates.

N.3. PR Crash Selection Bias Varies Inversely with Crash Severity Level

The effect size of the PR crash selection bias almost certainly varies by crash severity level, requiring individual adjustment for each crash severity level. It is well known that all fatal crashes in the US are uploaded to state and federal crash databases, so there is no PR crash selection bias for fatal crashes. However, only a certain (unknown) percentage of injury crashes with PARs are uploaded by police departments to their state crash database. It is plausible that an even smaller percentage of Major PDO crashes are uploaded than injury crashes. Finally, many local police departments may not upload Minor PDO crashes at all. Unfortunately, the percentages of uploads for injury, Major PDO, and Minor PDO crashes with PARs were not investigated in any of the studies reviewed in Appendix E. These previous studies mainly concerned unreported crashes rather than police-reported crashes that were unrecorded.

N.4. Effect on Google AV Standardized Crash Rates

Because of PR crash selection bias, the "SHRP2 6 Sites CV PR crash rate" (Var. #1 in Table H.1) is biased high relative to the "Mountain View CV police-recorded crash rate" (Var. #2). In turn, the "Standardization factor" (Var. #3) is biased high because its numerator is formed by the upwardly biased PR crash rate (Var. #1). Next, the estimate for the "Standard Google AV crashes" (Var. #5) is biased high because it is multiplied by the upwardly biased standardization factor (Var. # 3). Therefore, the "Standard Google AV crash rate" (Var. #7) is biased high because its numerator is the upwardly biased "Standard Google AV crashes" (Var. #5).

The effect size of the PR crash selection bias almost certainly varies with the crash severity level (Section N.3). Hence, the Google AV standardized Severe crash rate of 1.41 in Table H.1 (Var. #7) is the least upwardly biased crash rate of the three Severity Levels. Despite the upward bias, the Google AV standardized Severe crash rate of 1.41 is still below the SHRP2 CV Severe crash rate of 2.5 (see Section 3.4.1 and Figure 3.5, leftmost bars). That result gives rise to the protective Google AV Severe crash standardized IRR of 0.57 (Figure 3.6, leftmost square). Hence, the result shown in Figure 3.5 (that Google AVs reduce Severe crash rates compared to CVs) is valid, but the reduction would be even larger if the PR crash selection bias could be removed.

The Major PDO standardized crash rate of 3.5 (Table 3.2 and Figure 3.5, middle bars) is predicted to be even more upwardly biased than the Severe standardized crash rate. The reason is that an even lower percentage of Major PDO crashes than injury crashes are likely entered as police-recorded crashes in state databases. The Minor PDO standardized crash rate of 12.28 is predicted to be the most upwardly biased from PR crash selection bias. Therefore, the hypothesis cannot be excluded that Google AVs also reduce the Major PDO and Minor PDO crash rates compared to CVs. This hypothesis can be tested if the PR crash selection bias could be removed (see Section N.6).

N.5. Discussion

The standardized Google AV crash rate estimates in Table 3.2 and Figure 3.5 represent the worst case for Google AV crash rates. These AV crash rate estimates are what would occur if the local police departments at the SHRP2 sites had entered 100% of the PR crashes into state crash databases to become police-recorded crashes, but that only happens for fatal crashes. When fewer than 100% of the PR crashes become police-recorded crashes, the SHRP2 standardized crash rates would be even lower than the estimates given in Table 3.2 and plotted in Figure 3.5 (green-striped bars). As a result, the Google AV standardized crash IRR estimates are likely lower than the estimates given in Table 3.2 and plotted in Figure 3.6 (squares).

N.6. Future Research Recommendation

No known research study has investigated the percentage of PAR crashes entered (or not) into state or federal crash databases to make them police-recorded crashes. In particular, the percentage of PAR crashes that are not uploaded to become police-recorded crashes is unknown for Mountain View and unknown for the six SHRP2 sites. Hence, the PR crash selection bias cannot be removed based on known data.

Two methods are recommended to collect the data needed to adjust for PR crash selection bias:

1. The 46 SHRP2 CV crashes known to have a PAR in the Blanco et al. (2016) study [27] could be matched in state or federal crash databases to determine how many of the 46 PAR crashes were police-recorded at each crash Severity Level.
2. There were 323 police-recorded CV crashes on Mountain View city streets in 2015 (Table K.1, cell b) in the SWITRS database [28], all of which must have had a PAR. But with the cooperation of the Mountain View police department, it could be determined how many PARs at each crash Severity Level were not uploaded to become a police-recorded crash in the SWITRS database [28]. Similarly, the percentage of police-reported crashes with bench reports only (with no PAR) could also be determined for each crash Severity Level.

It is predicted that the higher the crash Severity Level,

1. A higher percentage of bench reports are converted to PARs.
2. A higher percentage of PARs are converted to police-recorded crashes.

Appendix Bibliography

References

1. U.S. Department of Transportation, "Preparing for the Future of Transportation: Automated Vehicles 3.0," Washington, DC, 2018, https://www.transportation.gov/av/3, accessed May 19, 2021.

2. Waymo, "Waymo Safety Report: On the Road to Fully Self-Driving," Mountain View, CA, 2018, https://storage.googleapis.com/sdc-prod/v1/safety-report/waymo-safety-report-2017.pdf, accessed June 29, 2021.

3. GM, "2018 Self-Driving Safety Report," General Motors, Detroit, MI, 2018, https://www.gm.com/content/dam/company/docs/us/en/gmcom/gmsafetyreport.pdf, accessed June 28, 2021.

4. Vander Wilt, N., "Aristotelian Logic Checker," Web Program, 2006, http://extinguishedscholar.com/temp/bob5.html, accessed May 19, 2021.

5. NHTSA, "National Motor Vehicle Crash Causation Survey," DOT HS 811 059, U.S. Department of Transportation, Washington, DC, July 2008, https://crashstats.nhtsa.dot.gov/Api/Public/ViewPublication/811059, accessed June 28, 2021; Singh, S., "Critical Reasons for Crashes Investigated in the National Motor Vehicle Crash Causation Survey," DOT HS 812 115, National Highway Traffic Safety Administration, Washington DC, 2015, http://www-nrd.nhtsa.dot.gov/Pubs/812115.pdf, accessed June 28, 2021; Singh, S., "Critical Reasons for Crashes Investigated in the National Motor Vehicle Crash Causation Survey," DOT HS 812 506, National Highway Traffic Safety Administration, Washington DC, 2018, https://crashstats.nhtsa.dot.gov/Api/Public/ViewPublication/812506, accessed June 28, 2021.

6. Rothman, K., Greenland, S., and Lash, T., *Modern Epidemiology*, 3rd ed. (Philadelphia, PA: Lippincott Williams & Wilkins, 2008). Later edition: Lash, T., VanderWeele, T., Haneause, S., and Rothman, K., *Modern Epidemiology*, 4th ed. (Philadelphia, PA: Lippincott Williams & Wilkins, 2020).

7. Dunn, N., Hickman, J., Soccolich, S., and Hanowski, R., "Applying the Crash Trifecta Approach to SHRP 2 Data," Contract No.: 18-UI-065, Virginia Tech Transportation Institute, Blacksburg, VA, 2018, https://vtechworks.lib.vt.edu/handle/10919/82744, accessed May 19, 2021.

8. Noy, I.Y., Shinar, D., and Horrey, W.J., "Automated Driving: Safety Blind Spots," *Safety Science* 102 (2018): 68-78, doi:10.1016/j.ssci.2017.07.018, https://www.sciencedirect.com/science/article/pii/S0925753517304198?via%3Dihub, accessed May 19, 2021.

9. Young, R.A., "Revised Odds Ratio Estimates of Secondary Tasks: A Re-Analysis of the 100-Car Naturalistic Driving Study Data," SAE Technical Paper 2015-01-1387, 2015, doi:https://doi.org/10.4271/2015-01-1387, https://www.researchgate.net/publication/275353775, accessed May 19, 2021.

10. Young, R.A., "Removing Biases from Crash Odds Ratio Estimates of Secondary Tasks: A New Analysis of the SHRP 2 Naturalistic Driving Study Data," SAE Technical Paper 2017-01-1380, 2017, doi:https://doi.org/10.4271/2017-01-1380, https://www.researchgate.net/profile/Richard_Young9/publication/319690973, accessed May 19, 2021.

11. Koopman, P., Kane, A., and Black, J., "Credible Autonomy Safety Argumentation," Paper Presented at in *27th Safety-Critical Systems Symposium*, Bristol, U.K., February 2019, http://users.ece.cmu.edu/~koopman/pubs/Koopman19_SSS_CredibleSafetyArgumentation.pdf, accessed May 19, 2021.

12. Koopman, P., "The Heavy Tail Safety Ceiling," Paper Presented at in *SAE Automated and Connected Vehicle Systems Testing Symposium*, Society of Automotive Engineers, Greenville, SC, June 2018, https://users.ece.cmu.edu/~koopman/pubs/koopman18_heavy_tail_ceiling.pdf, accessed May 19, 2021.

13. Dixit, V.V., Chand, S., and Nair, D.J., "Autonomous Vehicles: Disengagements, Accidents and Reaction Times," *PLOS ONE* 11, no. 12 (2016): e0168054, doi:10.1371/journal.pone.0168054, https://doi.org/10.1371/journal.pone.0168054, accessed May 19, 2021.

14. Kalra, N. and Paddock, S.M.., "Driving to Safety: How Many Miles of Driving Would It Take to Demonstrate Autonomous Vehicle Reliability?," *Transportation Research Part A: Policy and Practice* 94 (2016): 182-93, doi:10.1016/j.tra.2016.09.010, https://www.researchgate.net/publication/308538761, accessed May 19, 2021.

15. Wikipedia, "Failure Model and Effects Analysis," 2021, https://en.wikipedia.org/wiki/Failure_mode_and_effects_analysis, accessed May 19, 2021.

16. Kalra, N. and Groves, D.G., The Enemy of Good: Estimating the Cost of Waiting for Nearly Perfect Automated Vehicles, Rand Corporation, Santa Monica, CA, 2017, https://www.rand.org/pubs/research_reports/RR2150.html, accessed May 19, 2021.

17. U.S. Food and Drug Administration, "Development & Approval Process: Drugs," https://www.fda.gov/drugs/development-approval-process-drugs, accessed May 19, 2021.

18. Amrhein, V., Greenland, S., and McShane, B., "Scientists Rise up against Statistical Significance," *Nature* 567 (2019): 305-307, doi:10.1038/d41586-019-00857-9, https://www.nature.com/articles/d41586-019-00857-9, accessed May 19, 2021.

19. Wasserstein, R.L. and Lazar, N.A., "The ASA's Statement on *p*-Values: Context, Process, and Purpose," *The American Statistician* 70, no. 2 (2016): 129-133, doi:10.1080/00031305.2016.1154108, https://amstat.tandfonline.com/doi/full/10.1080/00031305.2016.1154108, accessed May 19, 2021.

20. Young, R.A., "Epidemiological Study Designs Applied to Driving Safety: Definitions and Examples," SAE Technical Paper 2018-01-0496, 2018, doi:https://doi.org/10.4271/2018-01-0496, accessed May 19, 2021.

21. SAE, "Taxonomy and Definitions for Terms Related to Driving Automation Systems for On-Road Motor Vehicles," SAE J3016_202104, SAE International, Warrendale, PA, 2021, https://www.sae.org/standards/content/j3016_202104/, accessed May 19, 2021.

22. Banerjee, S.S., Jha, S., Cyriac, J., Kalbarczyk, Z.T. et al., "Hands Off the Wheel in Autonomous Vehicles?: A Systems Perspective on over a Million Miles of Field Data," Paper Presented at in *2018 48th Annual IEEE/IFIP International Conference on Dependable Systems and Networks (DSN)*, Luxembourg City, Luxembourg, June 25-28, 2018, http://ssbaner2.cs.illinois.edu/publications/dsn2018/Paper.pdf, accessed May 19, 2021.

23. Porta, M., *A Dictionary of Epidemiology*, 6th ed. (New York: Oxford University Press, 2014), 343 pp., http://irea.ir/files/site1/pages/dictionary.pdf, accessed May 19, 2021.

24. HPMS, "California Department of Transportation Highway Performance Monitoring System Annual Reports," 2009-2014, https://dot.ca.gov/programs/research-innovation-system-information/highway-performance-monitoring-system, accessed May 19, 2021.

25. Hwang, J.G. and Yang, M.-C., "An Optimality Theory for Mid p-Values in 2 x 2 Contingency Tables," *Statistica Sinica* 11, no. 3, (2001): 807-826, http://www3.stat.sinica.edu.tw/statistica/oldpdf/A11n313.pdf, accessed May 19, 2021.

26. Teoh, E.R. and Kidd, D.G., "Rage Against the Machine? Google's Self-Driving Cars versus Human Drivers," *Journal of Safety Research* 63 (2017): 57-60, https://www.researchgate.net/profile/Eric_Teoh/publication/317265345, accessed May 19, 2021.

27. Blanco, M., Atwood, J., Russell, S., Trimble, T. et al., "Automated Vehicle Crash Rate Comparison Using Naturalistic Data," Virginia Tech Transportation Institute, January 2016, https://vtechworks.lib.vt.edu/handle/10919/64420, accessed May 19, 2021.

28. SWITRS, "California Highway Patrol Statewide Integrated Traffic Records System," https://iswitrs.chp.ca.gov/Reports/jsp/index.jsp, accessed May 19, 2021.

29. TIMS, "Transportation Injury Mapping System," https://tims.berkeley.edu/tools/query/summary.php, accessed May 19, 2021.

30. Google Reports, "Google Self-Driving Car Project Monthly Reports," Mountain View, CA, May 2015–August 2016. Note 1: Monthly reports are no longer publicly available from Google. However, a few monthly reports have been reproduced on other websites and may be found with Google searches, such as May 2015: https://www.slideshare.net/LudovicP/google-selfdriving-car-project-monthly-report-may-2015. Note 2: A partial summary of early Google AV crashes and mileage is given at https://medium.com/silk-stories/google-cars-november-update-e6e5b9721eb4. Note 3: 19 Google Reports from May 2015 to November 2016 are available at https://github.com/wgetsnaps/ca-dmv-autonomous-vehicle-accidents/tree/master/google-pdfs.

31. Schoettle, B. and Sivak, M., "A Preliminary Analysis of Real-World Crashes Involving Self-Driving Vehicles," Technical Report No. UMTRI-2015-34, The University of Michigan Transportation Research Institute, Ann Arbor, MI, 2015, http://www.umich.edu/~umtriswt/PDF/UMTRI-2015-34.pdf, accessed May 19, 2021.

32. CA DMV, "Report of Traffic Collision Involving an Autonomous Vehicle (OL 316)," California Department of Motor Vehicles, 2014-2019, https://www.dmv.ca.gov/portal/uploads/2021/05/OL316.pdf, accessed August 14, 2021. Note: Collision reports before January 1, 2019, have been archived by the CA DMV and are available upon request to them. However, CA DMV AV collision reports from October 14, 2014, to May 25, 2017, are available at https://wgetsnaps.github.io/ca-dmv-autonomous-vehicle-accidents/.

33. Favarò, F.M., Nader, N., Eurich, S.O., Tripp, M. et al., "Examining Accident Reports Involving Autonomous Vehicles in California," *PLOS ONE* 12, no. 9 (2017): e0184952, doi:10.1371/journal.pone.0184952, http://journals.plos.org/plosone/article?id=10.1371/journal.pone.0184952.

34. Efrati, A., "Waymo Collision Shows Flaws in Self-Driving Car Tests," The Information, October 2, 2018, https://www.theinformation.com/articles/waymo-collision-shows-flaws-in-self-driving-car-tests, accessed May 19, 2021.

35. Hall, S., "Google's Self-Driving Car Was Today in What Appears to Be Its Worst Accident Yet," September 23, 2016, https://9to5google.com/2016/09/23/googles-self-driving-car-was-today-in-what-appears-to-be-its-worst-accident-yet/, accessed May 19, 2021.

36. Blincoe, L., Miller, T.R., Zaloshnja, E., and Lawrence, B.A., "The Economic and Societal Impact of Motor Vehicle Crashes, 2010 (Revised)," Report No. DOT HS 812013, National Highway Traffic Safety Administration, Washington, DC, May 2015, https://trid.trb.org/view/1311862, accessed May 23, 2021.

37. Greenblatt, J., Merrin, M.B., Morganstein, D.E., and Schwartz, S., "National Accident Sampling System Nonreported Accident Survey," Report No. DOT HS 806 198, National Highway Traffic Safety Administration, Washington, DC, November 1981, https://trid.trb.org/view/213745, accessed May 19, 2021.

38. NHTSA, "National Telephone Survey of Reported and Unreported Motor Vehicle Crashes," Project No. DTNH22-08-C-0065, Findings Report, M. Davis & Co., National Highway Traffic Safety Administration, Washington, DC, 2011, https://crashstats.nhtsa.dot.gov/Api/Public/ViewPublication/812183, accessed May 19, 2021

39. M. Davis & Co., "National Telephone Survey of Reported and Unreported Motor Vehicle Crashes," Report No. DOT HS-812-183, National Highway Traffic Safety Administration, Washington, DC, July 2015. https://crashstats.nhtsa.dot.gov/Api/Public/ViewPublication/812183, accessed May 19, 2021.

40. Dingus, T.A., Klauer, S., Neale, V., Petersen, A. et al., The 100-Car Naturalistic Driving Study, Phase II - Results of the 100-Car Field Experiment, Virginia Tech Transportation Institute, 2006, http://trid.trb.org/view.aspx?id=783477.

41. CBS News, "Google Driverless Car Involved in First Injury-Causing Accident," July 17, 2015, https://www.cbsnews.com/news/google-driverless-car-involved-in-first-injury-causing-accident/, accessed May 19, 2021.

42. Mountain View Police Department, "Annual Reports, 2009-2018," https://www.mountainview.gov/google/asearch.asp?addsearch=police+department+annual+reports, accessed May 19, 2021.

43. Google Disengagements, 2015, 2016, "Google Self-Driving Car Testing Report on Disengagements of Autonomous Mode," Google, Inc., Mountain View, CA, https://www.consumerwatchdog.org/resources/cadmvdisengagereport-dec.2015.pdf and https://www.dmv.ca.gov/portal/vehicle-industry-services/autonomous-vehicles/disengagement-reports/, accessed May 19, 2021.

44. Blincoe, L.J. and Faigin, B.M., "The Economic Cost of Motor Vehicle Crashes, 1990," DOT HS 807 806, National Highway Traffic Safety Administration, U.S. Department of Transportation, Washington, DC, September 1992, https://www.google.com/books/edition/The_Economic_Cost_of_Motor_Vehicle_Crash/xfwdHeGmvmIC?hl=en&gbpv=1&dq=The+economic+cost+of+motor+vehicle+crashes+1990&printsec=frontcover, accessed June 30, 2021.

45. Blincoe, L.J., The Economic Cost of Motor Vehicle Crashes, 1994," DOT HS 808 425, National Highway Traffic Safety Administration, U.S. Department of Transportation, Washington, DC, July 1996, https://crashstats.nhtsa.dot.gov/Api/Public/ViewPublication/808425, accessed May 19, 2021.

46. Blincoe, L.J., Seay, A.G., Zaloshnja, E., Miller, T.R. et al., "The Economic Impact of Motor Vehicle Crashes, 2000," DOT HS 809 446, National Highway Traffic Safety Administration, Washington, DC, May 2002, https://crashstats.nhtsa.dot.gov/Api/Public/ViewPublication/809446, accessed May 19, 2021.

47. Bhattacharjee, R., "VIDEO: Google Self-Driving SUV Sideswipes Bus Near Company's Mountain View Headquarters," March 10, 2016, https://www.nbcsandiego.com/news/national-international/2021031/, accessed May 19, 2021.

48. Kokalitcheva, K., "Humans Cause Most Self-Driving Car Accidents," *Axios*, 2017 [updated August 3, 2017], https://www.axios.com/humans-cause-most-self-driving-car-accidents-1513304490-02cdaf3d-551f-46e6-ad98-637e6ef2c0b9.html, accessed May 19, 2021.

49. Baron, E., "Blame Game: Self-Driving Car Crash Highlights Tricky Legal Question," *The Mercury News*, January 25, 2018, https://www.mercurynews.com/2018/01/23/motorcyclist-hit-by-self-driving-car-in-s-f-sues-general-motors/, accessed May 19, 2021.

50. Felton, R., "GM Settles Lawsuit with Motorcyclist over Crash with Self-Driving Chevy Bolt," Jalopnik, June 1, 2018, https://jalopnik.com/gm-settles-lawsuit-with-motorcyclist-over-crash-with-se-1826492276, accessed May 19, 2021.

51. LaFrance, A., "When Google Self-Driving Cars Are in Accidents, Humans Are to Blame," *The Atlantic*, 2015, https://www.theatlantic.com/technology/archive/2015/06/every-single-time-a-google-self-driving-car-crashed-a-human-was-to-blame/395183/, accessed May 19, 2021.

52. Insurance Information Institute, "Background On: No-Fault Auto Insurance," 2018, https://www.iii.org/article/background-on-no-fault-auto-insurance, accessed May 19, 2021.

53. Rothman, K.J., "Episheet: Spreadsheets for the Analysis of Epidemiologic Data," 2015, https://www.colleaga.org/tools/episheet, accessed May 19, 2021.

54. CA DMV, "Vehicle Definitions," 2021, https://www.dmv.ca.gov/portal/dmv/detail/online/fee_calc/vehdef, accessed May 19, 2021.

55. Yue, L., Abdel-Aty, M., Wu, Y., and Wang, L., "Assessment of the Safety Benefits of Vehicles' Advanced Driver Assistance, Connectivity and Low Level Automation Systems," *Accident Analysis & Prevention* 117 (2018): 55-64, doi:10.1016/j.aap.2018.04.002, https://www.researchgate.net/publication/324531299, accessed May 19, 2021.

56. NHTSA, "Traffic Safety Facts: 2013 Data," DOT HS 812 16, National Highway Traffic Safety Administration, U.S. Department of Transportation, Washington, DC, 2015, https://crashstats.nhtsa.dot.gov/Api/Public/ViewPublication/812169, accessed May 19, 2021.

57. NHTSA, "2015 Motor Vehicle Crashes: Overview. Traffic Safety Facts: Research Note," DOT HS 812 318, National Highway Traffic Safety Administration, U.S. Department of Transportation, Washington, DC, 2016, https://crashstats.nhtsa.dot.gov/Api/Public/ViewPublication/812318, accessed May 19, 2021.

58. FHWA Federal Highway Administration, "Traffic Volume Trends – December 2015," U.S. Department of Transportation, Washington, DC, 2015, https://www.fhwa.dot.gov/policyinformation/travel_monitoring/15dectvt/15dectvt.pdf, accessed May 19, 2021.

59. StatsDirect, Statistical Software, Version 3.2.8, September 16, 2019, https://www.statsdirect.com/, accessed September 22, 2021.

60. CA DMV, California Code of Regulations Title 13, Division 1, Chapter 1, Article 3.7—Testing of Autonomous Vehicles, 2019, https://www.dmv.ca.gov/portal/uploads/2020/06/Adopted-Regulatory-Text-2019.pdf, accessed May 19, 2021.

Epilogue

This book reviewed and critically analyzed the first six scientific studies of AV crashes (Chapters 2, 3, 4, 5, 6, and 7). The AV-to-human crash rate ratio was generally overestimated in these studies. The book identified issues such as errors, biases, and best practice violations that caused these overestimates. The primary objective was to help future AV safety studies avoid these issues when comparing AV to CV crash rates.

During the reanalysis of the original data, I confirmed that AVs reduce Severe crashes, as first shown by Blanco et al. (2016).[1] In Section 3.4.1 (Chapter 3), I standardized the Google AV data of Blanco et al. (2016)[1] to adjust for site selection bias. The standardized data confirmed that prototype Google AVs reduced the Severe crash rate by 43% compared to CVs during the same period at the same site. In addition, I noticed that there had been no fatal crash in any prototype AV from any company in the CA field tests up to the time of writing the manuscript. These two results together predict that AVs may reduce fatal as well as Severe crashes relative to CVs, a substantial AV safety benefit if the prediction could be validated with additional data.

To test this fatal crash prediction, I extended the epidemiologic analysis methods in this book to fatal AV crash rates, comparing them to CV fatal crash rates.[2] All fatal crashes are recorded by the police in the United States, unlike non-fatal crashes, avoiding unrecorded crash selection bias (see Appendix E). That is, fatal CV crashes have no unrecorded crash selection bias, unlike non-fatal CV crashes. Another reason to test this prediction was that a prior study had claimed that "billions" of AV miles were required to validate AV fatal crash rates as lower than CV fatal crash rates.[3] My new study[2] tested this prediction by analyzing the fatal crash and mileage data from five companies with an AV fleet (Waymo, GM Cruise, Uber, Tesla, and Volvo). I proposed a new AV fatal crash safety validation criterion that "The upper confidence limit (UL) of the AV fatal crash rate has to be lower than the CV fatal crash rate with 95% confidence."[2] I found that "...epidemiologic analysis of real-world data finds that only millions of real-world AV [mileage] data are typically required to validate AV safety with 95% confidence to a 'Vision Zero' fatal crash criterion."[2] In fact, the AV mileage of some companies was already nearly large enough to meet the required AV miles to validate their zero or low fatal crash rates to this proposed criterion.[2]

[1] Blanco, M., Atwood, J., Russell, S., Trimble, T. et al., "Automated Vehicle Crash Rate Comparison Using Naturalistic Data," Virginia Tech Transportation Institute, January 2016, https://vtechworks.lib.vt.edu/handle/10919/64420, accessed May 19, 2021.

[2] Young, R., "Automated Driving System Safety: Miles for 95% Confidence in 'Vision Zero'," *SAE Int. J. Adv. & Curr. Prac. in Mobility* 2, no. 6 (2020): 3454-3480, https://doi.org/10.4271/2020-01-1205, accessed May 19, 2021.

[3] Kalra, N. and Paddock, S.M., "Driving to Safety: How Many Miles of Driving Would It Take to Demonstrate Autonomous Vehicle Reliability?," *Transportation Research Part A: Policy and Practice* 94 (2016): 182-193, doi:https://doi.org/10.1016/j.tra.2016.09.010, http://www.sciencedirect.com/science/article/pii/S0965856416302129, accessed May 19, 2021.

Hence, the Chapter 3 result that AVs reduce Severe crashes compared to CVs was supported and extended by this new analysis of AV fatal crashes.[2]

It is hoped that newer data from the CA DMV from 2017 and onwards will be examined in future studies. Hopefully, using methods that avoid the issues identified in this book. It can then be determined whether the more recent AV test data validate the AV IRR estimates and the crash rate trends identified in the 2014–2016 AV data examined in this book.

Analyses of AV crash data continue to be an active area of research. After completing this manuscript, several papers analyzed additional CA AV data (e.g., Wang and Li, 2019[4]; Biever et al., 2019[5]; Goodall, 2021[6]). Two other generic government reports were also published,[7,8] but had no empirical data. Khattak et al. (2020)[9] studied the relationship between AV disengagements and crashes using much of the same data analyzed in this book. They used a nested logistic model but somehow did not notice the substantial negative correlation between AV cumulative disengagement rates and AV cumulative crash rates seen in Figures 2.2C and 5.2C. Boogs et al. (2020)[10] asserted that "Automated vehicles (AVs) can reduce crashes by reducing or eliminating driver error," but they presented no evidence to support that assertion. A review article by Sohrabi et al. (2021)[11] summarized much of the AV safety literature to date. The article had a section on "Road test data analysis" that cited some of the same articles reviewed in this book. However, the results were just summarized and accepted at face value without the critical review of this book.

---

[4] Wang, S. and Li, Z., "Exploring the Mechanism of Crashes with Automated Vehicles Using Statistical Modeling Approaches," *PloS One* 14, no. 3 (2019): e0214550-e0214550, doi:10.1371/journal.pone.0214550, https://doi.org/10.1371/journal.pone.0214550, accessed August 14, 2021.

[5] Biever, W., Angell, L., and Seaman, S., "Automated Driving System Collisions: Early Lessons," *Human Factors* 62, no. 2 (2020): 249-259, https://www.researchgate.net/publication/335733786, accessed August 14, 2021.

[6] Goodall, N.J., "Comparison of Automated Vehicle Struck-from-Behind Crash Rates with National Rates Using Naturalistic Data," *Accident Analysis & Prevention* 154 (2021): 106056, https://www.sciencedirect.com/science/article/abs/pii/S0001457521000877, accessed August 14, 2021. For data see https://osf.io/eb4nu/.

[7] National Science and Technology Council and U.S. Department of Transportation, "Ensuring American Leadership in Automated Vehicle Technologies: Automated Vehicles 4.0," Washington, DC, January 2020, https://www.transportation.gov/sites/dot.gov/files/2020-02/EnsuringAmericanLeadershipAVTech4.pdf, accessed June 28, 2021.

[8] USDOT, "USDOT's Automated Vehicles Comprehensive Plan," January 2021, https://www.transportation.gov/sites/dot.gov/files/2021-01/USDOT_AVCP.pdf, accessed June 28, 2021.

[9] Khattak, Z.H., Fontaine, M.D. and Smith, B.L., "Exploratory Investigation of Disengagements and Crashes in Autonomous Vehicles under Mixed Traffic: An Endogenous Switching Regime Framework," *IEEE Transactions on Intelligent Transportation Systems:1-11*, 2020, doi:10.1109/TITS.2020.3003527, https://ieeexplore.ieee.org/document/9142420, accessed August 14, 2021.

[10] Boggs, A.M., Wali, B. and Khattak, A.J., "Exploratory Analysis of Automated Vehicle Crashes in California: A Text Analytics & Hierarchical Bayesian Heterogeneity-Based Approach," *Accident Analysis & Prevention* 135:105354, 2020/02/01/2020, doi:https://doi.org/10.1016/j.aap.2019.105354, https://www.sciencedirect.com/science/article/pii/S0001457519308735, accessed August 14, 2021.

[11] Sohrabi, S., Khodadadi, A., Mousavi, S.M., Dadashova, B. et al., "Quantifying the Automated Vehicle Safety Performance: A Scoping Review of the Literature, Evaluation of Methods, and Directions for Future Research," *Accid Anal Prev* 152:106003, Mar 2021, doi:10.1016/j.aap.2021.106003, https://www.sciencedirect.com/science/article/abs/pii/S0001457521000348?via%3Dihub, accessed August 14, 2021.

About the Author

Richard A. Young is President of Driving Safety Consulting, LLC, in Troy, Michigan. He has published over 70 papers, 17 with SAE. He is active on the following SAE committees: Safety and Human Factors Standards Steering Committee; Driver Metrics, Performance, Behaviors and States; Driver Vehicle Interface; Guidelines for Voice Interaction in Driver Vehicle Interface; and Driving Automation Systems.

Until June 2016, he was a Research Professor of Psychiatry and Cognitive Neuroscience at Wayne State University in Detroit, Michigan, with a joint appointment in the School of Medicine and the College of Engineering. He assumed those positions after retiring from General Motors (GM) in 2009 after 25 years.

Dr. Young received his PhD degree in experimental and physiological psychology with highest honors from New York University. His PhD thesis focused on human color perception, with Professor Edgar E. Coons as Thesis Advisor and Professor Leon Festinger as Research Advisor. He then received a Harvard Fellow award to investigate the neural mechanisms of learning with Professor Richard Thompson. He next accepted a Postdoctoral Fellowship at the University of California at Berkeley with Professor Russell De Valois. At Berkeley, Dr. Young did single-unit recordings in the monkey's visual system to investigate the neural mechanisms underlying color and form vision. He continued that research at the University of Oregon with a joint research appointment in the Psychology and Biology Departments. Dr. Young then accepted an offer from Apple Computer in technical support, where he supported Steve Jobs in the world press introduction of the Lisa computer in 1984. He then joined GM in the Computer Science Department at GM R&D, where he developed robot vision systems for 15 years based on his knowledge of the neural mechanisms of human vision. Next, he transferred to the Human Factors Department in GM Engineering, where he stayed for ten years, becoming GM's Global Lead in driver workload and driver vision.

Over the last 20 years, Dr. Young's primary research interests have been in the effects of secondary tasks on driver attention and performance, particularly cell phone conversation. He was the first to suggest that cell phone conversation does not increase crash risk and has done much research and published many journal articles concerning that topic. He has also recently served as an expert witness in legal cases concerning cell phone conversation, driver distraction, and crash risk.

Dr. Young began investigating autonomous vehicle safety. He found that many journal articles, company reports, and academic papers made safety claims (both positive and negative) about autonomous vehicles that contained systematic errors, biases, and best practice violations. This observation compelled him to write this book to improve the validity of future studies of autonomous vehicle safety.

Index

V

W

www.ingramcontent.com/pod-product-compliance
Lightning Source LLC
LaVergne TN
LVHW070909210926
848482LV00017B/1028
9781468603415